Progress Assessment
Support System
with Answer Key

Holt Social Studies

South and East Asia and the Pacific

HOLT, RINEHART AND WINSTON

A Harcourt Education Company

Orlando • **Austin** • New York • San Diego • Toronto • London

ISBN 0-03-078628-2

3 4 5 6 7 18 08 07

Contents

Smart Test: Test-Taking Tips for Students

TAKE IT ALL IN

When you finally hear the words "You may begin," briefly preview the test to get a mental map of your tasks:

- Know how many questions you have to complete.

- Set your time checkpoints.

STUDY THE DIRECTIONS

In order to follow directions, you have to know what the directions are! Read all the directions very carefully. Then study the answer sheet. How is it laid out? Be very sure you know exactly what to do and how to do it before you make the first mark.

I'M STUCK!

If you come across a question that stumps you, circle it and go on to others. Go back to the problem question later. What if you still have no idea? Use the 50/50 strategy and *make an educated guess*. Read every choice carefully. Use these pointers:

- *Watch out for distracters*—choices that may be true, but are too broad, too narrow, or not relevant to the question.

- Eliminate the least likely choice. Then, eliminate the next least likely choice, and so on until you find the best choice.

- If two choices seem equally correct, look to see if "all of the above" is an option. If it is, that might be the best choice. If no choice seems correct, look for "none of the above."

SEARCH FOR SKIPS AND SMUDGES

To avoid losing points on a machine graded test make sure you

- did not skip any answers

- gave one answer for each question

- made the marks heavy and dark and within the lines

- cleanly erased any smudges or stray pencil marks

NOTHING BUT THE TRUTH

Remember, it is sometimes easy to get tripped up on true/false questions. To avoid this, read the entire question before answering true or false. *The entire sentence or statement must be true if the answer is true.* If any part of the statement is false, the entire answer is false.

GETTING THE FULL PICTURE

When a question refers to a map, chart, or graph, read *all* the information carefully, including headings and labels. Determine any trends or oddities *before* answering the question.

I'M DONE!

Whoa! You aren't finished with your test until you check it. Take a look at how much time you have left. Go back and check for any careless mistakes you may have made, such as leaving a question blank or putting two answers to one question. Review the hardest questions, and turn the test in at the end of the time period.

READ AND RE-READ

If a question refers to a passage of text, read the selection, master the question, and then re-read the selection. The answer will probably pop out the second time around.

ANTICIPATE THE ANSWERS

Before you read the answer choices, *answer the question yourself*. If the answer you gave is among the choices listed, it is probably correct!

LOOK ALL AROUND

If the test item asks for vocabulary knowledge, remember to look at the surrounding sentences, or *context*, to see which definition fits. To identify the best definition of a word as it is used in the context, consider the surrounding words and phrases and information in nearby sentences.

NEGATIVE DO NOT FIT

Be sure to watch for negative words in questions such as *never*, *unless*, *not*, and *except*. If a question contains one of these words, look for the answer that *does not fit* with the other answers.

MASTER THE QUESTION

Have you ever said, "I knew the answer, but I thought the question asked something else"? Be very sure that you *know what a question is asking*. Read the question at least twice before reading the answer choices. Approach it as you would a mystery or a riddle. Look for clues.

South and East Asia and the Pacific

Diagnostic Test

MULTIPLE CHOICE For each of the following, write the letter of the best choice in the space provided.

______ **1.** Both Harappa and Mohenjo Daro were located near the
  a. Arabian Sea.
  b. Thar Desert.
  c. Chang Jiang River.
  d. Indus River.

______ **2.** With which of the following statements would the Buddha likely disagree?
  a. Suffering comes from not having what one wants.
  b. Nirvana is achieved by overcoming ignorance and desire.
  c. Contentment springs from gaining what one wants.
  d. Suffering and unhappiness are a part of human life.

______ **3.** What happened after the Guptas took control of India?
  a. Hinduism became more popular.
  b. Hinduism became less popular.
  c. Indians could not practice Buddhism.
  d. Indians could not practice Jainism.

______ **4.** The dynasty during which the Mongols ruled China was the
  a. Khan.
  b. Yuan.
  c. Song.
  d. Han.

______ **5.** Which of the following best describes the Qin dynasty under Shi Huangdi?
  a. His policies led to rebellion and civil war.
  b. His weak leadership resulted in the Warring States period.
  c. China lost territory under his rule.
  d. His strict policies kept China unified.

______ **6.** Which of the following correctly shows the order of dynasties in China?
  a. Sui, Song, Tang
  b. Sui, Tang, Song
  c. Tang, Song, Sui
  d. Song, Sui, Tang

______ **7.** What is the dominant religion of India today?
  a. Islam
  b. Christianity
  c. Hinduism
  d. Buddhism

______ **8.** What did the British agree to in 1947?
  a. independence for Bangladesh
  b. freeing Ghandi from prison
  c. ending the caste system
  d. the partition of India

Progress Assessment

_______ **9.** For much of the 20th century Mongolia was under the influence of
   a. Japan.
   b. Taiwan.
   c. the Soviet Union.
   d. China.

_______ **10.** Most Chinese live in the
   a. Taklimakan Desert.
   b. North China Plain.
   c. Plateau of Tibet.
   d. Sichuan Basin.

_______ **11.** What religion or belief system was brought to Korea and later carried to Japan by Chinese missionaries?
   a. Shinto
   b. Buddhism
   c. Confucianism
   d. Christianity

_______ **12.** Compared to California, Japan is
   a. half as large, with twice the population.
   b. twice as large, with half the population.
   c. about the same size, with four times the population.
   d. four times as large, with about the same population.

_______ **13.** What causes tsunamis?
   a. monsoons and typhoons
   b. underwater earthquakes and volcanic eruptions
   c. melting glaciers
   d. deforestation and flooding

_______ **14.** According to the domino theory
   a. if one country fell to communism, its neighbors would also.
   b. fighting for independence always results in a divided country.
   c. the only way for a country to achieve peace is with the help of the United Nations.
   d. a monarch may serve as head of state, but the legislature must hold the real power.

_______ **15.** What type of climate does most of Antarctica have?
   a. ice cap
   b. marine
   c. steppe
   d. highland

_______ **16.** Who were the first humans to live in Australia?
   a. Aborigines
   b. Maori
   c. Dutch settlers
   d. British prisoners

PRACTICING SOCIAL STUDIES SKILLS Study the map below and
answer the question that follows.

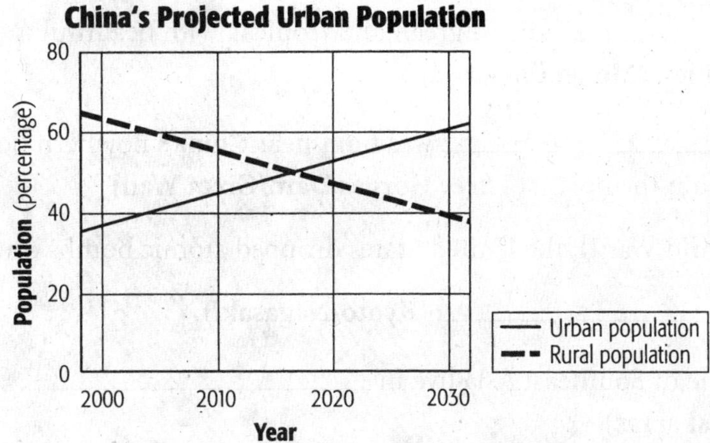

______ **1.** According to the graph, in about what year will the percentage of China's
urban and rural population be the same?

    a. 2009

    b. 2017

    c. 2023

    d. 2030

FILL IN THE BLANK Read each sentence and fill in the blank with the
word in the word pair that best completes the sentence.

1. The _____________________ arrived in the Indus River Valley from the
northwest. (**Harrapans/Aryans**)

2. Gupta society _____________________ after the rule of Candra Gupta II.
(**began to decline/adopted Hinduism**)

3. _____________________ taught proper behavior, and also emphasized spiritual
matters. (**Confucianism/Neo-Confucianism**)

4. Leaders of the Song dynasty used the _____________________ examination to
make sure qualified people became government workers.
(**scholar officials/civil service**)

5. _____________________ are seasonal winds that bring either moist or dry air
to an area. (**Savannas/Monsoons**)

6. The ___________________ was a program in India to encourage farmers to adopt modern farming methods. (**green revolution/partition**)

7. ___________________ is green and tropical, with beautiful mountains and crowded cities. (**Mongolia/Taiwan**)

8. The ___________________ was built near China's northern border to stop invaders from the north. (**Three Gorges Dam/Great Wall**)

9. During World War II, the United States dropped atomic bombs on Hiroshima and ___________________. (**Kyoto/Nagasaki**)

10. Most people of Southeast Asia live in ___________________. (**cities/rural areas**)

11. Mining is an important industry in the ___________________, Australia's interior. (**Great Barrier Reef/Outback**)

12. A ___________________ is a high latitude region that receives little precipitation. (**polar desert/marine**)

TRUE/FALSE Indicate whether each statement below is true or false by writing **T** or **F** in the space provided.

______ **1.** The caste system divided Indian society into groups based on a person's birth, wealth, or occupation.

______ **2.** During the Zhou dynasty, people believed their rulers had a mandate of heaven.

______ **3.** Mohandas Ghandi was a main opponent of the Indian independence movement.

______ **4.** Millions of people in India have moved out of cities in search of jobs.

______ **5.** Taiwan is an island nation, while Mongolia is landlocked.

______ **6.** China welcomed contact with the outside world throughout much of its history.

______ **7.** Korea's main obstacle to reunification is the question of a common form of government.

______ **8.** Japan has many diverse ethnic groups that all speak different languages.

______ **9.** Southeast Asia lies in the tropics, the area around the Arctic Circle.

______ **10.** Singapore is one of the world's busiest free ports.

______ **11.** The Antarctic Treaty of 1959 designated the continent as an approved site for oil drilling and military activity.

______ **12.** Raising livestock is important to the economies of Australia and New Zealand.

MATCHING In the space provided, write the letter of the term or place that matches each description. Some answers will not be used.

______ **1.** people who want to spread their religious beliefs

______ **2.** most heavily populated Pacific Island region

______ **3.** a landform at the mouth of a river created by sediment deposits

______ **4.** the most important language of ancient India

______ **5.** a group of nomadic peoples who invaded China

______ **6.** a large group of islands

______ **7.** the world's coldest desert

______ **8.** fees a country charges on exports and imports

______ **9.** territory claimed by both India and Pakistan

______ **10.** Japan's highest mountain

a. Gobi

b. Melanesia

c. Mongols

d. wats

e. delta

f. Kashmir

g. tariffs

h. Polynesia

i. missionaries

j. archipelago

k. Sanskrit

l. Fuji

History of Ancient India

Section Quiz

Section 1

MULTIPLE CHOICE For each of the following, write the letter of the best choice in the space provided.

_______ **1.** The Harappan civilization developed in the valley of which river?
  a. the Inga
  b. the Tigris
  c. the Indus
  d. the Nile

_______ **2.** The city of Mohenjo Daro had which of the following features?
  a. indoor plumbing
  b. electric lights
  c. Greek architecture
  d. Buddhist temples

_______ **3.** What items have archaeologists found that help us understand Harappan society?
  a. writings
  b. statues
  c. carts with wheels
  d. paintings

_______ **4.** What were Aryan religious writings called?
  a. Vedas
  b. rajas
  c. Sanskrit
  d. Ganges

_______ **5.** In what language were the sacred hymns and poems of the Aryans composed?
  a. Persian
  b. Hindi
  c. Aryan
  d. Sanskrit

Progress Assessment

History of Ancient India

Section Quiz

Section 2

FILL IN THE BLANK For each of the following statements, fill in the blank with the appropriate word, phrase, or name.

1. A system of social classes arose in India with the arrival of the

   _______________________, whose sacred texts described four main classes

   in society.

2. The religion practiced by the Aryans was called _______________________.

3. The oldest of the four collections of hymns and poems sacred to the Aryans is the

   _______________________.

4. The _______________________ further divided Indian society into groups based on a person's rank, wealth, or occupation.

5. _______________________, the largest religion in India today, is a blending of many of the religious traditions of Central Asia and the Indian subcontinent.

6. _______________________, the process of rebirth, is a central belief of Hinduism.

7. Hindus believe that each god is part of a single universal spirit called

   _______________________.

8. _______________________ is the effects that good or bad actions have on a person's soul.

9. _______________________ was based on the teachings of a man named Mahavira.

10. The Jains practice _______________________ to fulfill one of the main principles of their faith, to injure no life.

Progress Assessment

History of Ancient India

Section Quiz

Section 3

MATCHING In the space provided, write the letter of the term or person that matches each description. Some answers will not be used.

______ 1. the prince who questioned the meaning of human life

______ 2. going without food

______ 3. focusing of the mind on spiritual ideas

______ 4. "Enlightened One"

______ 5. the guiding principles at the heart of the Buddha's teachings

______ 6. a state of perfect peace

______ 7. method for overcoming ignorance and desire

______ 8. people who work to spread their religious beliefs

______ 9. people who tried to follow the Buddha's teachings exactly as they had been stated

______ 10. believed that other people could interpret the Buddha's teachings

a. Brahmins

b. Buddhism

c. Eightfold Path

d. fasting

e. Four Noble Truths

f. Hinduism

g. Mahayana Buddhists

h. meditation

i. missionaries

j. nirvana

k. reincarnation

l. Siddhartha Gautama

m. the Buddha

n. Theravada Buddhists

Progress Assessment

History of Ancient India

Section Quiz

Section 4

TRUE/FALSE Mark each statement **T** if it is true or **F** if it is false. If false explain why.

______ **1.** In the 320s BC a military leader named Candragupta Maurya seized control of the entire southern part of India.

____

____

______ **2.** Asoka was a strong ruler, the strongest of all the Mauryan emperors.

____

____

______ **3.** Under the rule of Candra Gupta II, Indian society continued to grow, the economy strengthened, and people prospered.

____

____

______ **4.** Rulers of the Gupta dynasty generally were not supportive of the Hindu caste system.

____

____

______ **5.** India was firmly under Gupta control until the late 400s, when Jainist armies invaded India from the northwest.

____

____

Progress Assessment

History of Ancient India

Section Quiz

Section 5

FILL IN THE BLANK For each of the following statements, fill in the blank with the appropriate word, phrase, or name.

1. Indian artists of the Maurya and Gupta periods created great works of art, often of a _________________ nature.

2. Most of the literature created during the Maurya and Gupta periods was written in _________________.

3. The ancient Indians were masters of _________________, the science of working with metals.

4. Indian doctors understood the concept of _________________, which helps build human defenses to a disease.

5. The *Mahabharata,* one of the longest literary works in the world, contains many passages about _________________ beliefs.

6. The _________________ is the most famous of the passages from the *Mahabharata.*

7. Demonstrating the importance of religion in ancient Indian society, many of the finest works of art from the period can be found in _________________.

8. An _________________ is a mixture of two or more metals.

9. The numerals we use today were created by _________________ scholars and brought to Europe by Arabs.

10. Indian scholars were skilled in _________________, which enabled them to identify seven of the nine planets.

Progress Assessment

History of Ancient India

Chapter Test

Form A

MULTIPLE CHOICE For each of the following, write the letter of the best choice in the space provided.

______ **1.** Both Harappa and Mohenjo Daro were located near the
  a. city of Bodh Gaya.
  b. Arabian Sea.
  c. Thar Desert.
  d. Indus River.

______ **2.** Which of the following statements describes the Aryan system of government?
  a. It was based on a strong central government.
  b. It did not allow local leaders.
  c. It was led by scholars who could write.
  d. Its leaders were often skilled warriors.

______ **3.** In Aryan society, which of the following was of highest importance in regard to the caste system?
  a. moving to a higher level of the caste system
  b. interacting with people of different castes
  c. adhering to the strict rules of the caste system
  d. helping people from lower castes to rise up

______ **4.** Which of the following *best* describes the relationship between the Vedic texts and the Vedas?
  a. Vedic texts were a collection of thoughts about the Vedas.
  b. Vedic texts contained the same sacred hymns as the Vedas.
  c. The Vedas were based on the teachings in the Vedic texts.
  d. The Vedas described how Aryans should use the Vedic texts.

______ **5.** Which of the following events happened last?
  a. Siddhartha Gautama wandered through forests for years.
  b. Siddhartha Gautama became disenchanted with life.
  c. Siddhartha Gautama gained insight into human suffering.
  d. Siddhartha Gautama meditated underneath a tree.

Progress Assessment

______ **6.** According to Hinduism, people are reborn into new physical forms. The form one is born into depends on one's
  a. moksha.
  b. karma.
  c. dharma.
  d. sutra.

______ **7.** With which of the following statements would the Buddha likely disagree?
  a. Suffering comes from not having what one wants.
  b. Contentment springs from gaining what one wants.
  c. Nirvana is achieved by overcoming ignorance and desire.
  d. Suffering and unhappiness are a part of human life.

______ **8.** Which of the following describes Theravada Buddhism?
  a. Theravada Buddhists follow the Buddha's teachings exactly.
  b. Theravada Buddhists have rejected the Buddha's teachings.
  c. Theravada Buddhism is the largest branch of Buddhism today.
  d. Theravada Buddhism is based on the teachings of Mahariva.

______ **9.** What happened after the Guptas took control of India?
  a. Hinduism became more popular.
  b. Hinduism became less popular.
  c. Indians could not practice Buddhism.
  d. Indians could not practice Jainism.

______ **10.** Which of the following describes how Hindu temples changed in the Gupta period?
  a. The temples became more basic and simplistic.
  b. The temples were built without towers.
  c. The temples no longer included carvings of gods.
  d. The temples became more complex.

______ **11.** Which Indian piece of literature contains the passages called the *Bhagavad Gita*?
  a. the *Mahabharata*
  b. the *Siddheswara*
  c. the *Ramayana*
  d. the *Panchatantra*

PRACTICING SOCIAL STUDIES SKILLS Study the map below and answer the question that follows.

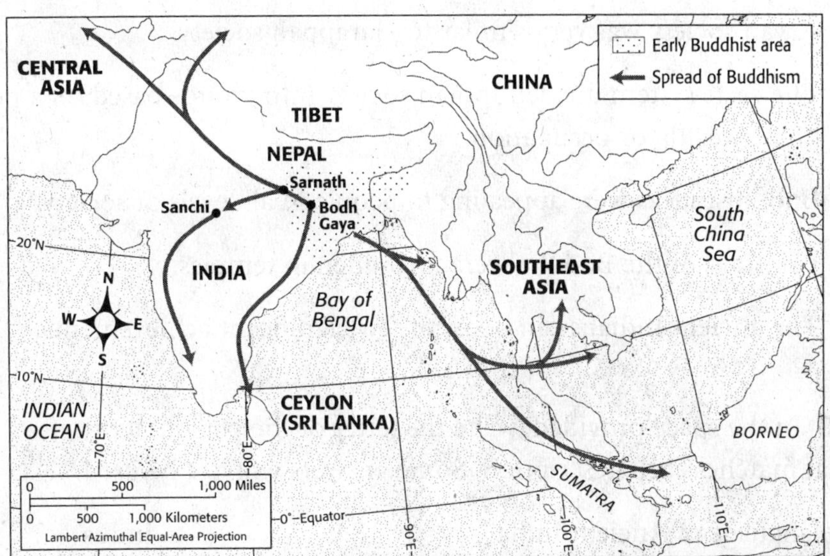

_______ **1.** About how far did early Buddhist missionaries have to travel to reach Ceylon?
　　a. less than 250 miles
　　b. about 500 miles
　　c. about 750 miles
　　d. more than 1200 miles

FILL IN THE BLANK Read each sentence and fill in the blank with the word in the word pair that best completes the sentence.

1. The ___________________ arrived in the Indus valley from the northwest.
　　(Harrapans/Aryans)

2. ___________________ was based on the teaching of a man named Mahariva.
　　(Buddhism/Jainism)

3. A person's ___________________ determined his or her place in society.
　　(Veda/caste)

4. Gupta society began to decline after the rule of ___________________.
　　(Candra Gupta II/Candragupta Maurya)

5. Ancient Indians mixed metals to create ___________________.
　　(alloys/inoculations)

TRUE/FALSE Indicate whether each statement below is true or false by writing **T** or **F** in the space provided.

_______ **1.** Aryan society was very similar to Harappan society.

_______ **2.** The caste system divided Indian society into groups based on a person's birth, wealth, or occupation.

_______ **3.** Buddhist ideas were appealing to people in all levels of society.

_______ **4.** Paintings of the Buddha were forbidden in temples.

_______ **5.** The ancient Indians can boast of few significant achievements.

MATCHING In the space provided, write the letter of the person, term, or place that matches each description. Some answers will not be used.

_______ **1.** language of Ancient India

_______ **2.** a major river flowing through India

_______ **3.** the division of Indian society into groups

_______ **4.** the process of a soul being reborn in a new body

_______ **5.** the force created by a person's actions

_______ **6.** going without food

_______ **7.** people who spread their religious beliefs

_______ **8.** a military leader who seized control of northern India

_______ **9.** the science of working with metals

_______ **10.** injecting a person with a small dose of a virus

a. Indus

b. inoculation

c. caste system

d. reincarnation

e. metallurgy

f. Candragupta Maurya

g. fasting

h. missionaries

i. Sanskrit

j. Vedic

k. karma

History of Ancient India

Chapter Test

Form B

SHORT ANSWER Answer each of the following questions in complete sentences. Remember to use specific examples to support your answers.

1. In what ways did the geography of India influence the development of civilizations?

____

____

2. What did the Buddha teach about material goods?

____

____

3. What were some of the major accomplishments of the Gupta period?

____

____

PRACTICING SOCIAL STUDIES SKILLS Study the map below and answer the question that follows.

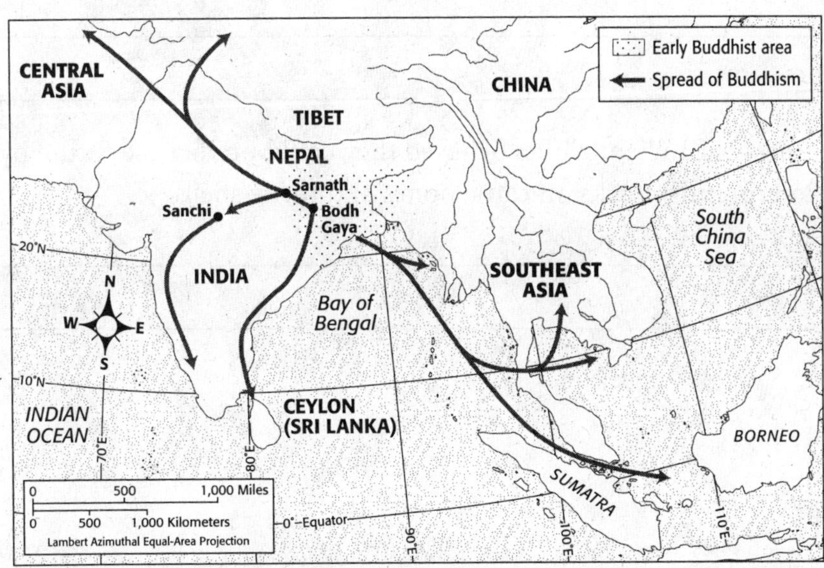

1. Would you expect to find more Buddhists in India or Sumatra? Explain your answer.

____

____

 Progress Assessment

History of Ancient China

Section Quiz

Section 1

TRUE/FALSE Mark each statement **T** if it is true or **F** if it is false. If false explain why.

_______ **1.** China's physical geography made the land unsuitable for farming, therefore early civilizations relied on the sea for food.

____

____

_______ **2.** Civilization in China began along the Huang He and Chang Jiang rivers.

____

____

_______ **3.** The Qin dynasty lasted longer than any other Chinese dynasty.

____

____

_______ **4.** The Zhou believed that no one could rule China without permission from heaven.

____

____

_______ **5.** Priests of ancient China believed they could predict the future by interpreting cracks on cattle bones and turtle shells.

____

____

Progress Assessment

History of Ancient China

Section Quiz

Section 2

FILL IN THE BLANK For each of the following statements, fill in the blank with the appropriate word, phrase, or name.

1. Emperor _____________________ made Confucianism China's official government philosophy.

2. The second class in Han China, the largest, was made up of the

   _____________________.

3. One item that the Han Chinese made that we use every day was

   _____________________.

4. To get a government job in China under Han rule, a person had to pass a test based on the teachings of _____________________.

5. _____________________ was the first emperor of the Han dynasty.

6. Confucius taught that the _____________________ was the head of the family and had absolute power within the family.

7. The _____________________, a device that measures the strength of an earthquake, was invented during the Han dynasty.

8. Emperors were very interested in knowing the movements of the earth because they believed that _____________________ were signs of future evil events.

9. A _____________________ uses the position of shadows cast by the sun to tell the time of day.

10. _____________________ is the practice of inserting fine needles through the skin at specific points to cure disease or relieve pain.

Progress Assessment

History of Ancient China

Section Quiz

Section 3

MATCHING In the space provided, write the letter of the term, person, or place that matches each description. Some answers will not be used.

_______ **1.** ruler who reunified China and created the Sui dynasty

_______ **2.** China's capital and largest city of the Tang dynasty

_______ **3.** series of waterways that linked major cities

_______ **4.** a thin beautiful pottery invented by the Chinese

_______ **5.** an important Chinese export of which the method of making it was kept secret

_______ **6.** perhaps China's greatest female poet

_______ **7.** popular substitute for bulky coins made of metal

_______ **8.** manner by which the world's first known book was printed

_______ **9.** dramatically altered how wars were fought

_______ **10.** instrument that uses the earth's magnetic field to indicate direction

a. silk

b. Chang'an

c. compass

d. celadon

e. Empress Wu

f. fireworks

g. Grand Canal

h. gunpowder

i. Yang Jian

j. Li Qingzhao

k. movable type

l. paper money

m. porcelain

n. woodblock printing

Progress Assessment

History of Ancient China

Section Quiz

Section 4

FILL IN THE BLANK For each of the following statements, fill in the blank with the appropriate word, phrase, or name.

1. _______________________ is a philosophy based on the ideas of the ancient Chinese philosopher Confucius.

2. The teachings of Confucius focused on _______________________, or proper behavior, for individuals and governments.

3. Confucius thought that _______________________ was maintained when people knew their place and behaved appropriately.

4. Confucianism became the official state philosophy of the _______________________ dynasty.

5. As _______________________ became more popular in China, Confucianism lost some of its influence.

6. A _______________________ is a body of unelected government officials.

7. _______________________ means service as a government official.

8. Successfully passing the civil service examination led to a career as a _______________________, an educated member of the government.

9. _______________________ was similar to Confucianism but it also emphasized spiritual matters.

10. The civil service system was a major factor in the stability of the _______________________ dynasty government.

History of Ancient China

Section Quiz

Section 5

MATCHING In the space provided, write the letter of the term, person, or place that matches each description. Some answers will not be used.

_______ **1.** nomadic peoples from north of China who began their conquest of China in 1211

_______ **2.** leader who led his army into northern China in 1211

_______ **3.** self declared emperor of China who founded the Yuan dynasty in 1279

_______ **4.** dynasty during which emperors worked to eliminate foreign influences from Chinese society

_______ **5.** Italian merchant who traveled through China beginning in 1271 and 1295

_______ **6.** founder of the Ming dynasty who became emperor of China after defeating the Mongols in 1368

_______ **7.** a huge palace complex that included almost 1,000 buildings, all of which were off-limits to the common people

_______ **8.** greatest sailor of the period who led seven grand voyages between 1405 and 1433

_______ **9.** a policy of avoiding contact with other countries

_______ **10.** protected China from northern invaders

a. conquest

b. empire

c. Forbidden City

d. Genghis Khan

e. Great Wall

f. isolationism

g. kamikaze

h. Kublai Khan

i. Marco Polo

j. Ming

k. Mongols

l. trade

m. Zheng He

n. Zhu Yuanzhang

Progress Assessment

History of Ancient China　　　　　　　　Chapter Test

Form A

MULTIPLE CHOICE For each of the following, write the letter of the best choice in the space provided.

______ **1.** The Shang were the first people in China to
 a. use advanced metal tools.
 b. develop a writing system.
 c. hunt with bows and arrows.
 d. domesticate pigs and sheep.

______ **2.** Which of the following best describes the Qin dynasty?
 a. The strict policies of Shi Huangdi led to rebellion and civil war.
 b. The weak leadership of Shi Huangdi resulted in the Warring States period.
 c. China lost territory under Shi Huangdi's rule.
 d. The strict policies of Shi Huangdi kept China unified.

______ **3.** Compare Liu Bang to all of the other Chinese leaders before him. In which way was he different?
 a. He was the longest-serving emperor of China.
 b. He was the first student of Confucius to become emperor.
 c. He was the only Daoist teacher to become emperor.
 d. He was the first common person to become emperor.

______ **4.** Which Chinese invention is used to measure the strength of earthquakes?
 a. sundial
 b. acupunture
 c. porcelain
 d. seismograph

______ **5.** During the Han dynasty, Chinese painters became experts at
 a. using mosaics.
 b. making paper.
 c. painting figures.
 d. sketching buildings.

______ **6.** Which of the following correctly shows the order of dynasties in China?
 a. Sui, Song, Tang
 b. Sui, Tang, Song
 c. Tang, Song, Sui
 d. Song, Sui, Tang

　　　　　　　　　　　　　　　　Progress Assessment

_____ **7.** What effect did opening the Pacific ports to foreign traders have on China?

    a. It gave foreign empires easy access routes to invade China.

    b. China's export business collapsed because most goods were brought in from elsewhere.

    c. It expanded trading, contributing to a strong economy in China.

    d. A great number of foreigners moved to these Chinese port cities.

_____ **8.** What effect did having a bureaucracy of scholar-officials have on the Song dynasty?

    a. It created stability and an efficient government.

    b. It created a government in which nothing was accomplished because of mismanagement.

    c. It created sharp divisions within the society, eventually leading to a civil war.

    d. It created a very unstable government.

_____ **9.** The dynasty during which the Mongols ruled China was the

    a. Khan.

    b. Yuan.

    c. Song.

    d. Han.

_____ **10.** Which of the following best characterizes Genghis Khan's expeditions of conquest?

    a. closely fought battles

    b. mostly peaceful conquests in which villages either surrendered or retreated

    c. slow, drawn-out attacks, often taking months

    d. bloody attacks in which entire populations of cities and towns were often wiped out

                    Progress Assessment

PRACTICING SOCIAL STUDIES SKILLS Study the map below and
answer the question that follows.

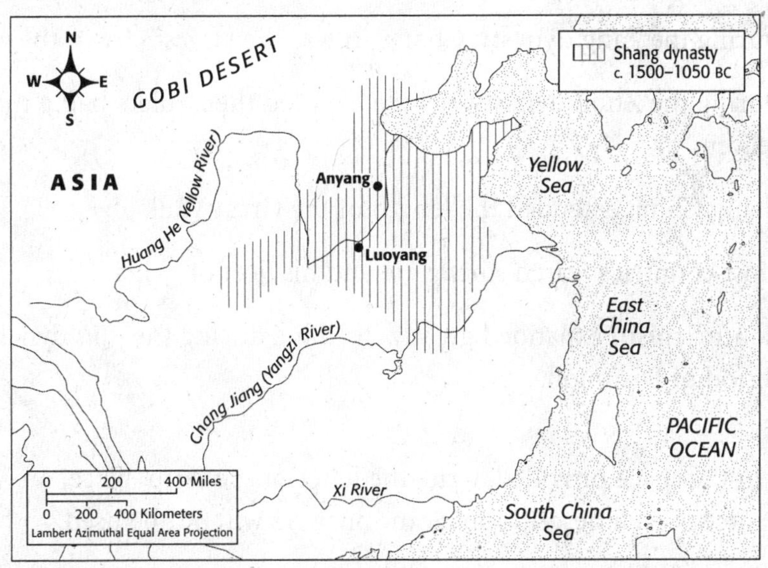

_______ **1.** What natural feature formed part of the southern border of Shang China?
   a. Japan
   b. Gobi Desert
   c. Chang Jiang
   d. Huang He

FILL IN THE BLANK Read each sentence and fill in the blank with the
word in the word pair that best completes the sentence.

   1. During the Shang dynasty the first _________________ in China was
   invented. (**writing/paper**)

   2. _________________ taught proper behavior, and also emphasized
   spiritual matters. (**Confucianism/Neo-Confucianism**)

   3. The _________________ was more than 2,000 miles long and was located
   near China's northern border. (**Grand Canal/Great Wall**)

   4. Han artwork showed _________________. (**daily activities/gods and myths**)

   5. Song dynasty leaders created the _________________ examination to make
   sure qualified people became government workers. (**scholar-officials/civil service**)

                                      Progress Assessment

TRUE/FALSE Indicate whether each statement below is true or false by writing T or F in the space provided.

_______ **1.** During the Tang dynasty, Chang'an was the largest city in the world.

_______ **2.** During the Zhou dynasty, people believed their rulers had a mandate of heaven.

_______ **3.** Shang China extended the length of the Great Wall.

_______ **4.** Most countries feared Zheng He and his fleet of ships.

_______ **5.** Confucianism expanded greatly in China during the Qin dynasty and the Period of Disunion.

MATCHING In the space provided, write the letter of the term, place, or person that matches each description. Some answers will not be used.

_______ **1.** a thin, beautiful type of pottery

_______ **2.** leader of the Mongols

_______ **3.** people who made up most of the population in Han China

_______ **4.** person whose teachings focused on ethics

_______ **5.** a device used to tell the time of day

_______ **6.** the strong Qin king who unified China

_______ **7.** the only woman to rule China

_______ **8.** a group of nomadic peoples who attacked China

_______ **9.** Chinese sailor who led many voyages

_______ **10.** the capital city of the Song dynasty

a. Shi Huangdi

b. sundial

c. Confucius

d. porcelain

e. Empress Wu

f. Zheng He

g. Chang'an

h. Kaifeng

i. Genghis Khan

j. acupuncture

k. compass

l. peasants

m. Mongols

History of Ancient China

Chapter Test

Form B

SHORT ANSWER Answer each of the following questions in complete sentences. Remember to use specific examples to support your answer.

1. Why was the Tang dynasty called the golden age of Chinese civilization?

____

____

2. How did Chinese society change under the Ming?

____

____

3. What kind of government did Wudi want to create?

____

____

PRACTICING SOCIAL STUDIES SKILLS Study the map below and answer the question that follows.

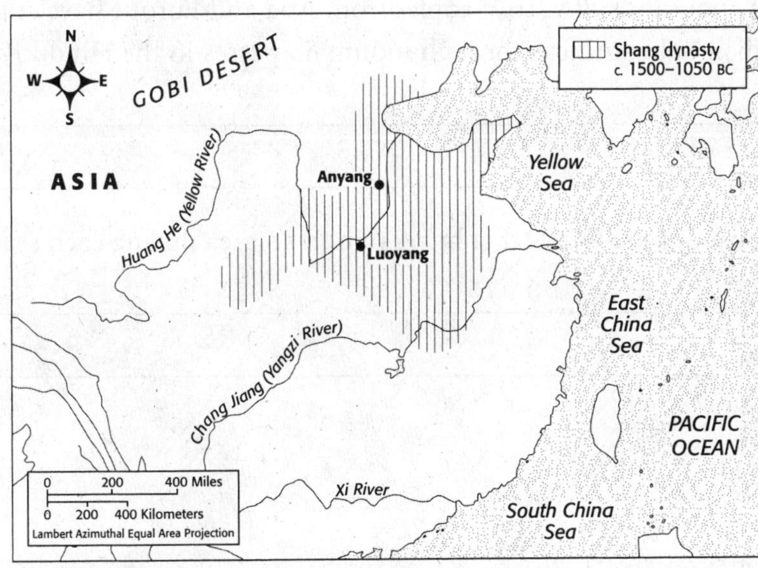

1. Explain how flooding might have been a major problem to the Shang dynasty.

____

____

The Indian Subcontinent

Section Quiz

Section 1

TRUE/FALSE Mark each statement **T** if it is true or **F** if it is false. If false explain why.

_______ **1.** The Indus River creates a huge delta in Bangladesh.

____

____

_______ **2.** The Indian Subcontinent has a variety of climates, but all of them are warm.

____

____

_______ **3.** The Himalayas stretch for about 1,500 miles along the northern border of the Indian Subcontinent.

____

____

_______ **4.** For thousands of years, peoples from Asia and Europe have entered the Indian Subcontinent through mountain passes in the Hindu Kush.

____

____

_______ **5.** Monsoons are winds that bring rain to an area during each season.

____

____

 Progress Assessment

The Indian Subcontinent

Section Quiz

Section 2

MATCHING In the space provided, write the letter of the term or person that matches each description. Some answers will not be used.

_______ **1.** divided Indian society into groups based on a person's birth or occupation

_______ **2.** the division of India

_______ **3.** a territory inhabited and controlled by people from a foreign land

_______ **4.** Indian independence movement leader

_______ **5.** religion of most people in Pakistan

_______ **6.** ruled India, starting in the 1800s until 1947

_______ **7.** language of the Aryans

_______ **8.** Muslim warrior who established the Mughal Empire

_______ **9.** in Buddhism, a state of perfect peace in which suffering ends

_______ **10.** the dominant religion in India

a. Sanskrit

b. nirvana

c. Babur

d. partition

e. Asoka

f. Islam

g. Great Britain

h. tolerance

i. colony

j. caste system

k. Hinduism

l. Mohandas Gandhi

m. Buddhism

Progress Assessment

The Indian Subcontinent

Section Quiz

Section 3

MULTIPLE CHOICE For each of the following, write the letter of the best choice in the space provided.

______ **1.** Since 1947, India's population has
  a. stayed the same.
  b. decreased slightly.
  c. doubled.
  d. reached 2 billion.

______ **2.** The increase in the percentage of people who live in India's cities is called
  a. population density.
  b. reverse migration.
  c. mobilization.
  d. urbanization.

______ **3.** The green revolution was a program to encourage
  a. the use of modern agricultural techniques.
  b. maintenance of healthy-looking lawns.
  c. replanting of forests that have been cut down.
  d. irrigation of the Thar Desert.

______ **4.** India's two largest cities are
  a. Mumbai and Bangalore.
  b. Kolkata and Mumbai.
  c. Delhi and Bangalore.
  d. Kolkata and Delhi.

______ **5.** The popular festival Diwali is also known as
  a. the day of the dead.
  b. the celebration of independence.
  c. new years's day.
  d. the festival of lights.

The Indian Subcontinent

Section Quiz

Section 4

FILL IN THE BLANK For each of the following statements, fill in the blank with the appropriate word, phrase, or name.

1. Kathmandu is the capital of _________________.

2. Dhaka is the capital of _________________.

3. In 2004, thousands of people in Sri Lanka were killed by a

    _________________.

4. India and Pakistan have clashed over the territory of _________________.

5. The _________________ of Nepal often serve as guides through the Himalayas.

 Progress Assessment

The Indian Subcontinent

Chapter Test

Form A

MULTIPLE CHOICE For each of the following, write the letter of the best choice in the space provided.

______ **1.** In 1971, East Pakistan broke off to become the independent country of
   a. Nepal.
   b. Bangladesh.
   c. Sri Lanka.
   d. Diwali.

______ **2.** The first urban civilization on the Indian Subcontinent was the
   a. Tamil.
   b. Aryan.
   c. Mauryan.
   d. Harappan.

______ **3.** The worldest highest mountain is
   a. K2.
   b. Mount Everest.
   c. Ghats.
   d. Hindu Kush.

______ **4.** Indian troops revolted against British rule in the
   a. 1500s.
   b. 1600s.
   c. 1800s.
   d. 1900s.

______ **5.** Indian language and culture were greatly influenced by the
   a. Sinhalese.
   b. Tamil.
   c. Sherpas.
   d. Aryans.

______ **6.** More than 70 percent of India's population lives
   a. on mountain farms.
   b. in the suburbs.
   c. in villages.
   d. in cities.

______ **7.** What country has a larger population than India?
   a. the United States
   b. Russia
   c. China
   d. Pakistan

______ **8.** The Indus River flows through
   a. India.
   b. Bangladesh.
   c. Sri Lanka.
   d. Pakistan.

______ **9.** What system divided Indian society into groups based on birth or occupation?
   a. partition
   b. caste
   c. colonies
   d. urbanization

______ **10.** Which Indian leader helped expand the Mauryan Empire?
   a. Asoka
   b. Babur
   c. Akbar
   d. Gandhi

Progress Assessment

_______ **11.** Under whose rule was the Taj Mahal built?
  a. Mughal
  b. Aryan
  c. Harappan
  e. Mauryan

_______ **12.** What did the British agree to in 1947?
  a. independence for Bangladesh
  b. freeing Ghandi from prison
  c. ending the caste system
  d. the partition of India

_______ **13.** The Buddha taught that people can rise above their desire for material goods and reach
  a. Brahman.
  b. tolerance.
  c. reincarnation.
  d. nirvana.

_______ **14.** One of the biggest challenges facing Bangladesh is
  a. war with Pakistan.
  b. flooding.
  c. ethnic conflict.
  d. drought.

_______ **15.** Which of these countries is an island?
  a. Sri Lanka
  b. Nepal
  c. Bhutan
  d. Kathmandu

_______ **16.** What is the dominant religion of India?
  a. Islam
  b. Hinduism
  c. Christianity
  d. Buddhism

_______ **17.** The Ganges River joins with other rivers and creates a huge
  a. tributary.
  b. delta.
  c. subcontinent.
  d. monsoon.

_______ **18.** The Mughal Empire grew rich from trade in goods such as
  a. oil and coal.
  b. gold and silver.
  c. rice and corn.
  d. tea and spices.

_______ **19.** In the struggle for Indian independence, Mohandas Gandhi used the strategy of
  a. urbanization.
  b. partition.
  c. nonviolent protest.
  d. guerrilla warfare.

_______ **20.** Which ethnic group in Nepal often serves as Himalayan guides?
  a. Tamil
  b. Bhutan
  c. Gupta
  d. Sherpas

PRACTICING SOCIAL STUDIES SKILLS Study the map below and answer the question that follows.

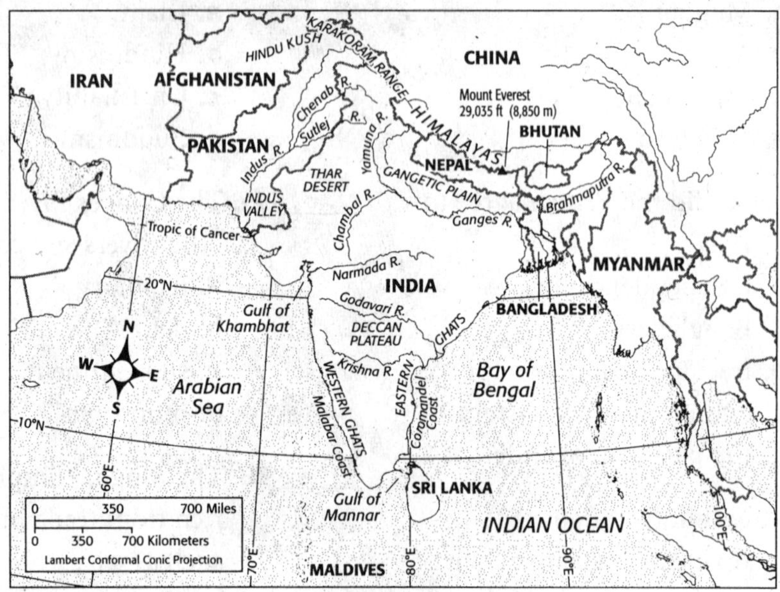

______ **1.** Which mountain range contains Mt. Everest?
   a. Karakoram Range
   b. Eastern Ghats
   c. Himalayas
   d. Hindu Kush

FILL IN THE BLANK Read each sentence and fill in the blank with the word in the word pair that best completes the sentence.

1. The Harappan civilization was founded around 2300 BC in the

_______________________ River Valley. **(Hindu Kush/Indus)**

2. The capital of Nepal is _______________________. **(Kathmandu/Dhaka)**

3. The beginnings of India's social system and Hindu religion came from the

_______________________. **(Aryans/British)**

4. _______________________ are seasonal winds that bring either moist or dry air
to an area. **(Savannas/Monsoons)**

5. The _______________________ was a program in India to encourage farmers to
adopt modern farming methods. **(green revolution/partition)**

The Indian Subcontinent, *continued*　　　　　　　　　Chapter Test Form A

TRUE/FALSE Indicate whether each statement below is true or false by writing **T** or **F** in the space provided.

______ **1.** Millions of people in India have moved out of cities in search of jobs, causing urbanization.

______ **2.** The Indian Subcontinent's most important resource is petroleum.

______ **3.** Bangladesh is one of the world's most densely populated countries.

______ **4.** Mohandas Ghandi was the most important leader of the Indian independence movement.

______ **5.** Most people in Pakistan practice Hinduism.

MATCHING In the space provided, write the letter of the term or place that matches each description. Some answers will not be used.

______ **1.** desert in northwestern India

______ **2.** in Buddhism, a state of perfect peace in which suffering ends

______ **3.** dominant religion in Bangladesh

______ **4.** small mountain kingdom in the Himalayas

______ **5.** a large landmass that is smaller than a continent

______ **6.** a territory controlled by a foreign land

______ **7.** Pakistan's most densely populated region

______ **8.** the dominant religion in both Bhutan and Sri Lanka

______ **9.** territory claimed by both Pakistan and India

______ **10.** a landform at the mouth of a river created by sediment deposits

a. delta

b. colony

c. Jainism

d. Buddhism

e. Thar

f. Kashmir

g. Bhutan

h. Islam

i. nirvana

j. subcontinent

k. partition

l. Indus River Valley

　　　　　　　　　　　　　　　　　　　　　Progress Assessment

The Indian Subcontinent

Chapter Test

Form B

SHORT ANSWER Answer each of the following questions in complete sentences. Remember to use specific examples to support your answers.

1. Why did 10 million people cross borders on the Indian Subcontinent in 1947?

____

____

2. What kind of climate does most of the Indian Subcontinent have?

____

____

3. What are the main challenges facing India?

____

____

PRACTICING SOCIAL STUDIES SKILLS Study the map below and answer the question that follows.

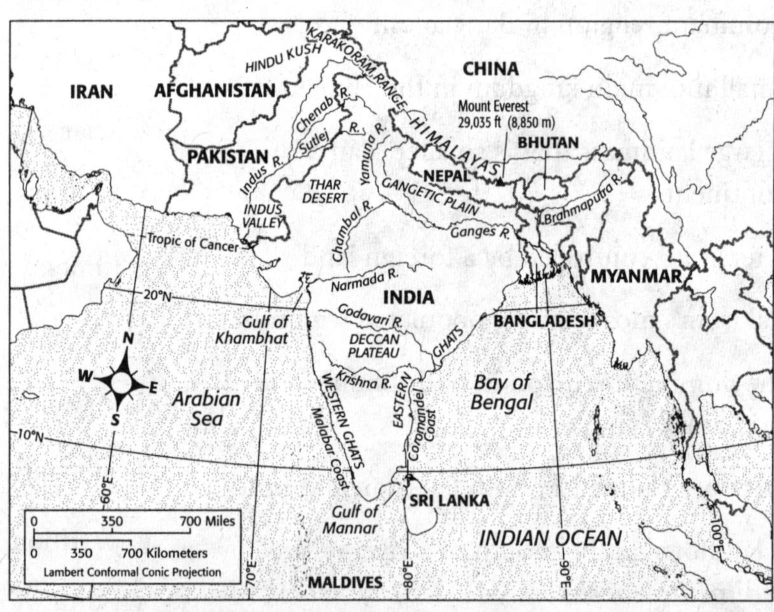

1. Why do you think the earliest civilizations on the Indian Subcontinent were located in the northern part of the region?

____

____

36

China, Mongolia, and Taiwan

Section Quiz

Section 1

MATCHING In the space provided, write the letter of the term or place that matches each description. Some answers will not be used.

_______ **1.** the world's highest mountains

_______ **2.** a major river in northern China

_______ **3.** located in Mongolia, the world's coldest desert

_______ **4.** fertile, yellowish soil

_______ **5.** storms that bring high winds and rain

_______ **6.** green tropical island

_______ **7.** Asia's longest river

_______ **8.** landlocked country

_______ **9.** the world's highest plateau

_______ **10.** main population center of China

a. Gobi

b. Himalayas

c. loess

d. North China Plain

e. typhoons

f. Chang Jiang

g. Plateau of Tibet

h. Huang He

i. Mongolia

j. Taklimakan

k. Mongolian Plateau

l. Taiwan

Progress Assessment

China, Mongolia, and Taiwan

Section Quiz

Section 2

FILL IN THE BLANK For each of the following statements, fill in the blank with the appropriate word, phrase, or name.

1. Most people in China belong to the _______________________ ethnic group.

2. After their defeat in the civil war, the Nationalists fled to

   _______________________.

3. The Qin was the first _______________________ to unite China under one empire.

4. _______________________ was a Communist leader during the civil war against the Nationalists.

5. Many Chinese blend elements of their religions with _______________________, a philosophy that stresses the importance of family, moral values, and respect for one's elders.

6. Most Chinese speak Mandarin or a regional _______________________.

7. Chinese art often includes _______________________, or decorative writing.

8. China's main belief systems today are Buddhism and _______________________, which stresses living simply and in harmony with nature.

9. Most Chinese live in the Manchurian and _______________________ plains.

10. _______________________ are Buddhist temples with multi-storied towers and upward curving roofs.

Progress Assessment

China, Mongolia, and Taiwan

Section Quiz

Section 3

MULTIPLE CHOICE For each of the following, write the letter of the best choice in the space provided.

_______ **1.** China's capital city is
  a. Shanghai.
  b. Beijing.
  c. Hong Kong.
  d. Macao.

_______ **2.** Which cities were European colonies until the late 1990s?
  a. Hong Kong and Macao
  b. Shanghai and Beijing
  c. Peking and Tiananmen Square
  d. Taiwan and Tibet

_______ **3.** What is the Three Gorges Dam intended to do?
  a. bring more water to urban dwellers
  b. protect plant and animal habitats
  c. generate electrical power
  d. encourage the expansion of farmland

_______ **4.** Many countries have considered limiting or stopping trade with China because it does not
  a. work to address pollution.
  b. have a large economy.
  c. allow privately owned businesses.
  d. respect human rights.

_______ **5.** Under a command economy, the government
  a. closes state-run factories.
  b. owns all businesses and makes all decisions.
  c. allows privately owned businesses.
  d. encourages foreign businesses to own companies.

China, Mongolia, and Taiwan

Section Quiz

Section 4

TRUE/FALSE Mark each statement **T** if it is true or **F** if it is false. If false explain why.

_______ **1.** When the Nationalists took over Taiwan, they established a Communist government.

____

____

_______ **2.** The Mongol Empire was once a great world power.

____

____

_______ **3.** Kao-hsiung is Taiwan's main seaport and center of heavy industry.

____

____

_______ **4.** Mongolia has been a democratic country since it declared independence from China.

____

____

_______ **5.** Mongolia has many large cities, including the capital, Ulaanbaatar.

____

____

China, Mongolia, and Taiwan

Chapter Test

Form A

MULTIPLE CHOICE For each of the following, write the letter of the best choice in the space provided.

______ **1.** Which religion or set of beliefs stresses living simply and in harmony with nature?
- **a.** Daoism
- **b.** Buddhism
- **c.** ancestor worship
- **d.** communism

______ **2.** The southeast is sometimes struck by violent storms called
- **a.** monsoons.
- **b.** typhoons.
- **c.** blizzards.
- **d.** sandstorms.

______ **3.** Most Mongolians make their living as
- **a.** horse racers.
- **b.** livestock herders.
- **c.** rice farmers.
- **d.** tourist guides.

______ **4.** China's first Communist leader was
- **a.** Mao Zedong.
- **b.** Shi Huangdi.
- **c.** Deng Xiaoping.
- **d.** Genghis Khan.

______ **5.** For much of the 20th century, Mongolia was under the influence of
- **a.** Japan.
- **b.** Taiwan.
- **c.** the Soviet Union.
- **d.** China.

______ **6.** China's last dynasty was the
- **a.** Qing.
- **b.** Shang.
- **c.** Beijing.
- **d.** Zhou.

______ **7.** Many countries consider China's response to rebellions as violations of
- **a.** trade agreements.
- **b.** human rights.
- **c.** free enterprise principles.
- **d.** environmental protections.

______ **8.** The most dominant influence on Taiwan's culture today is
- **a.** Chinese.
- **b.** Japanese.
- **c.** Mongolian.
- **d.** European.

China, Mongolia, and Taiwan, *continued*   Chapter Test Form A

______ **9.** Attempts to modernize China's economy after 1976 were begun by
  a. Chiang Kai-shek.
  b. Mao Zedong.
  c. Deng Xiaoping.
  d. the Dalai Lama.

______ **10.** What led to the events of Tiananmen Square in 1989?
  a. People rebelled in Tibet.
  b. Foreign businesses had begun to own Chinese companies.
  c. Protesters demanded political rights and freedoms.
  d. Countries threatened to limit or stop trade with China.

______ **11.** The Three Gorges Dam will create power from
  a. water.
  b. wind.
  c. gasoline.
  d. coal.

______ **12.** Most Chinese live in the
  a. Taklimakan Desert.
  b. North China Plain.
  c. Plateau of Tibet.
  d. Sichuan Basin.

______ **13.** China was ruled for centuries by
  a. the Soviet Union.
  b. Westerners.
  c. Japan.
  d. family dynasties.

______ **14.** Who are the Han?
  a. Tibetan leaders
  b. martial artists
  c. Portuguese traders
  d. China's largest ethnic group

______ **15.** China's largest city is
  a. Taipei.
  b. Shanghai.
  c. Beijing.
  d. Macao.

______ **16.** Taiwan's climate most resembles that of
  a. northern China.
  b. western Mongolia.
  c. central Tibet.
  d. southeastern China.

PRACTICING SOCIAL STUDIES SKILLS Study the graph below and
answer the question that follows.

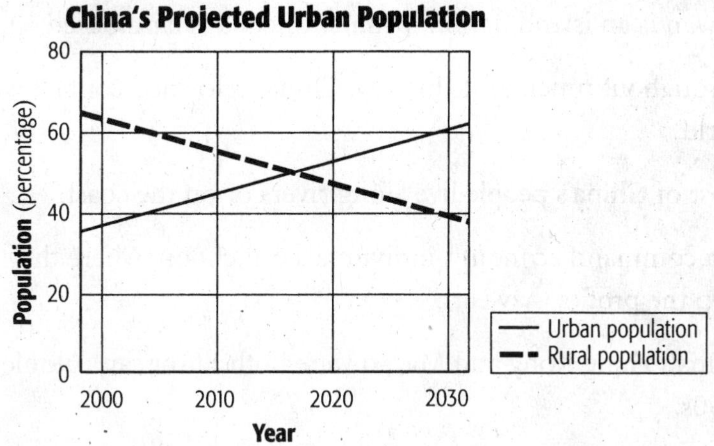

_______ **1.** About what percentage of China's population is expected to be urban
in 2030?
a. 20 percent
b. 40 percent
c. 60 percent
d. 80 percent

FILL IN THE BLANK Read each sentence and fill in the blank with the
word in the word pair that best completes the sentence.

1. China hopes that the _______________________ will lessen pollution by creating
more electrical power. (**Three Gorges Dam/*Ilha Formosa***)

2. The _______________________ was built in northern China to stop invaders
from the north. (**Great Wall/Forbidden City**)

3. About one fourth of Mongolia's people live in the capital city,

_______________________. (**Kao-hsiung/Ulaanbaatar**)

4. _______________________ is green and tropical, with beautiful mountains and
crowded cities. (**Mongolia/Taiwan**)

5. China's main belief systems are Buddhism and _______________________ .
(**Daoism/Islam**)

TRUE/FALSE Indicate whether each statement is true or false by writing
T or **F** in the space provided.

______ **1.** Taiwan is an island nation, while Mongolia is landlocked.

______ **2.** Throughout much of its history, China welcomed contact with the outside world.

______ **3.** Most of China's people live along rivers or on the coast.

______ **4.** In a command economy, individuals can choose where they work and keep the profits they earn.

______ **5.** China's Hong Kong and Macao were both European colonies until the late 1990s.

MATCHING In the space provided, write the letter of the term or
person that matches each description. Some answers will not be used.

______ **1.** a philosophy that stresses family, moral values, and respect for elders

______ **2.** a Nationalist leader

______ **3.** a river in northern China

______ **4.** the world's coldest desert

______ **5.** once under British control, a city on China's coast

______ **6.** an important Chinese dynasty that united China under one emperor

______ **7.** an economic system in which the government owns all the businesses and makes all the decisions

______ **8.** fertile, yellowish soil

______ **9.** the world's highest mountain range

______ **10.** a Mongolian ruler

a. Hong Kong

b. Huang He

c. loess

d. Genghis Khan

e. Confucianism

f. Chiang Kai-shek

g. Gobi

h. Himalayas

i. command economy

j. Qin

k. Mao Zedong

l. Zhou

China, Mongolia, and Taiwan

Chapter Test

Form B

SHORT ANSWER Answer each of the following questions in complete
sentences. Remember to use specific examples to support your answers.

1. How has China responded to those who have criticized or opposed the
   government? Give two examples.

   ____

   ____

2. Why is Mongolia sparsely populated?

   ____

   ____

3. Generally, how does the region's climate change from the southeast to the
   northwest?

   ____

   ____

PRACTICING SOCIAL STUDIES SKILLS Study the graph below and
answer the question that follows.

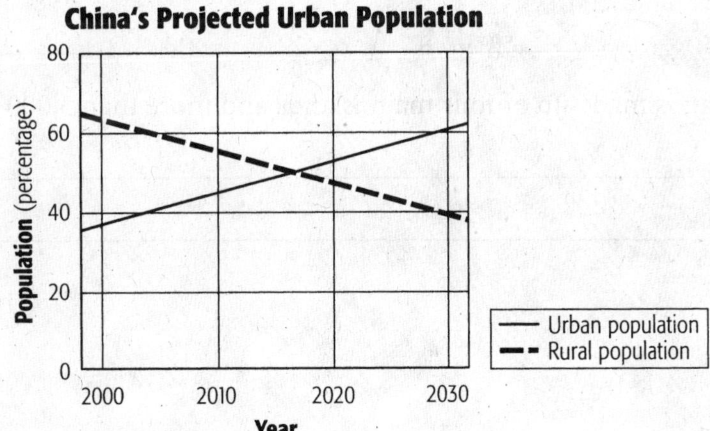

1. According to the graph, by how about much will the percentage of China's
   urban population rise between 2000 and 2030?

   ____

   ____

                                     Progress Assessment

Japan and the Koreas

Section Quiz

Section 1

TRUE/FALSE Mark each statement **T** if it is true or **F** if it is false. If false explain why.

______ **1.** North Korea has mineral resources such as coal and iron ore.

____

____

______ **2.** Underwater earthquakes sometimes cause destructive waves called typhoons.

____

____

______ **3.** Plains cover about 75 percent of Japan.

____

____

______ **4.** Unlike Japan, Korea has some large plains, which are found along its western coast.

____

____

______ **5.** Japan is made up of four main islands and more than 3,000 smaller ones.

____

____

Progress Assessment

Japan and the Koreas

Section Quiz

Section 2

MATCHING In the space provided, write the letter of the term or place that matches each description. Some answers will not be used.

_______ **1.** Korean dish made from pickled cabbage and spices

_______ **2.** traditional theater of Japan

_______ **3.** highly trained warrior

_______ **4.** influenced early history of Korea

_______ **5.** type of character used in Japanese writing system

_______ **6.** religion native to Japan

_______ **7.** traditional robe worn in Japan on special occasions

_______ **8.** imperial capital of Japan

_______ **9.** city devastated in World War II by an atomic bomb

_______ **10.** generals who ruled Japan in the emperor's name

a. kanji

b. Kyoto

c. samurai

d. China

e. kami

f. Buddhism

g. shoguns

h. kabuki

i. kimchi

j. Nagasaki

k. Shinto

l. kimono

Progress Assessment

Japan and the Koreas

Section Quiz

Section 3

MULTIPLE CHOICE For each of the following, write the letter of the best choice in the space provided.

______ **1.** Japan's main exports are
   a. foods such as rice and tea.
   b. manufactured goods.
   c. coal and iron ore.
   d. high speed trains.

______ **2.** Who makes the laws in Japan today?
   a. the emperor and the Diet
   b. the chancellor and the Congress
   c. the prime minister and the Diet
   d. the president and the Congress

______ **3.** Most farms in Japan are
   a. similar to farms in the United States.
   b. located high in the mountains.
   c. cooperatives owned by their members.
   d. too small for farmers to make a living.

______ **4.** Japan's second largest city is
   a. Kyoto.
   b. Osaka.
   c. Hirohito.
   d. Mitsubishi.

______ **5.** Most Japanese workers are
   a. hard working and loyal.
   b. poorly educated but hard working.
   c. highly trained and underpaid.
   d. hard working but disloyal.

Name _________________________________ Class _______________ Date _____________

Japan and the Koreas

Section Quiz

Section 4

FILL IN THE BLANK For each of the following statements, fill in the
blank with the appropriate word, phrase, or name.

1. _____________________ Korea is rich in mineral resources.

2. Since the end of World War II North and South Korea have been separated by a

   _____________________.

3. The official name of South Korea is the _____________________ of Korea.

4. North Korea is ruled by a dictator who is advised by members of the

   _____________________ Party.

5. The chief obstacle to the reunification of Korea is the disagreement over the form

   of _____________________.

6. _____________________ took over as the dictator of North Korea after the
   death of his father, Kim Il Sung.

7. South Korea's economy is one of the _____________________ in East Asia.

8. The South Korean government hopes to put an end to _____________________
   through reform programs.

9. Many countries worry about North Korea's ability to make

   _____________________ weapons.

10. The capital of South Korea is _____________________.

Progress Assessment

Japan and the Koreas

Chapter Test

Form A

MULTIPLE CHOICE For each of the following, write the letter of the best choice in the space provided.

______ **1.** What religion was brought to Korea and later carried to Japan by Chinese missionaries?
- **a.** Shinto
- **b.** Buddhism
- **c.** Confucianism
- **d.** Christianity

______ **2.** Most of Japan's major cities are located on the island of
- **a.** Honshu.
- **b.** Hokkaido.
- **c.** Shikoku.
- **d.** Kyushu.

______ **3.** About how many people live in Pyongyang?
- **a.** 300,000
- **b.** 1,000,000
- **c.** 3,000,000
- **d.** 10,000,000

______ **4.** The average farm in the U.S. is how many times as large as the average farm in Japan?
- **a.** 5
- **b.** 15
- **c.** 50
- **d.** 175

______ **5.** The Korean War began in
- **a.** 1941.
- **b.** 1950.
- **c.** 1963.
- **d.** 1994.

______ **6.** In Japan, the elected legislature is called the
- **a.** Kyoto Protocol.
- **b.** Diet.
- **c.** Parliament.
- **d.** Shogun.

______ **7.** What type of characters represent whole words in the Japanese writing system?
- **a.** Kanji
- **b.** Kami
- **c.** Kana
- **d.** Kimchi

______ **8.** Compared to California, Japan is
- **a.** half as large, with twice the population.
- **b.** twice as large, with half the population.
- **c.** about the same size, with four times the population.
- **d.** four times as large, with about the same population.

______ **9.** The Democratic People's Republic of Korea (North Korea) is really
- **a.** a republic.
- **b.** a democracy.
- **c.** a democracy and a republic.
- **d.** neither a democracy nor a republic.

 Progress Assessment

_______ **10.** Kimchi is made from pickled
    a. cabbage.
    b. eggs.
    c. fish.
    d. beets.

_______ **11.** About what percent of South Koreans are Christian?
    a. 5 percent
    b. 10 percent
    c. 25 percent
    d. 60 percent

_______ **12.** Japan's successful trade policy has created a
    a. large food supply.
    b. strong work ethic.
    c. huge trade surplus.
    d. shortage of raw materials.

_______ **13.** Powerful storms that affect both Japan and the Korean Peninsula are called
    a. monsoons.
    b. typhoons.
    c. cyclones.
    d. tsunamis.

_______ **14.** After World War II, North Korea formed a government with the aid of
    a. China.
    b. the United Nations.
    c. the United States.
    d. the Soviet Union.

_______ **15.** In the past, most people in Korea were Buddhist and
    a. Shintoist.
    b. Hindu.
    c. Christian.
    d. Confucianist.

_______ **16.** Unlike Japan, Korea has
    a. some large plains.
    b. rugged mountains.
    c. summer typhoons.
    d. destructive earthquakes.

_______ **17.** Some members of the families that control much of South Korea's industry have used their wealth and power to make themselves
    a. very popular.
    b. religious leaders.
    c. protectors of the poor.
    d. even wealthier.

_______ **18.** North Korean leaders discourage people from
    a. playing baseball.
    b. practicing religion.
    c. following old traditions.
    d. joining the Communist Party.

PRACTICING SOCIAL STUDIES SKILLS Study the map below and
answer the question that follows.

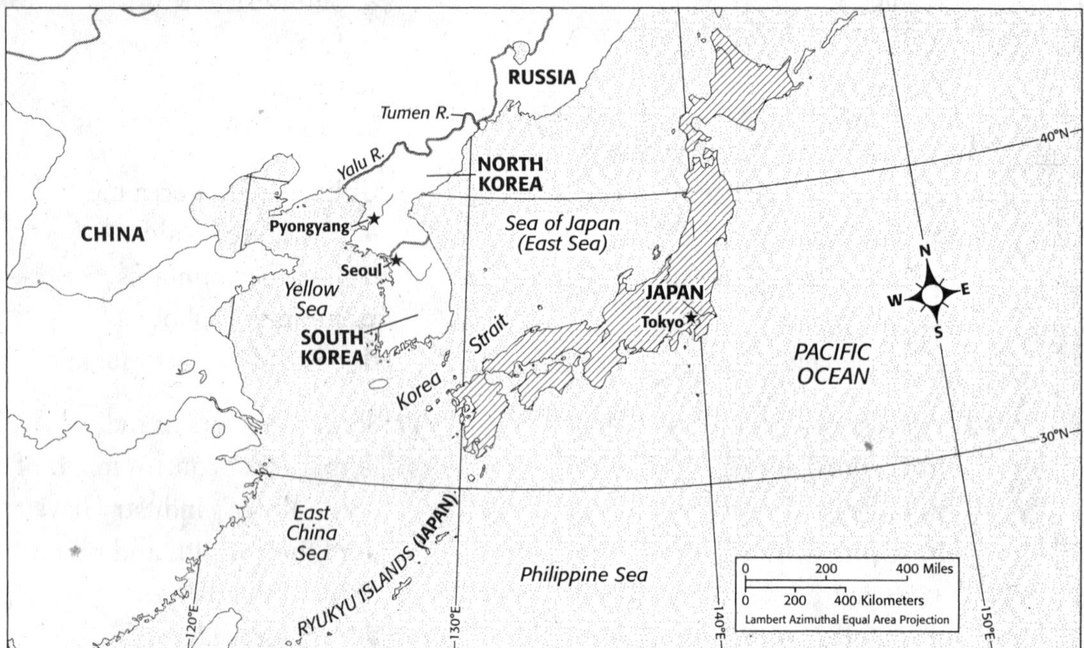

_______ **1.** About how far is it by air from Tokyo, Japan, to Seoul, South Korea?

    a. 525 miles

    b. 725 miles

    c. 850 miles

    d. 1150 miles

FILL IN THE BLANK Read each sentence and fill in the blank with the
word in the word pair that best completes the sentence.

1. Korea was ruled by _________________________ before it was ruled by Japan.
   (Kim Il Sung/China)

2. The United States dropped atomic bombs on Hiroshima and

_________________________. **(Nagasaki/Kyoto)**

3. Shoguns ruled Japan with the help of armies of _________________________.
   (samurai/generals)

4. Kim Jong Il is the ruler of _________________________.
   (North Korea/South Korea)

5. Japan's imperial capital was _________________________. **(Osaka/Kyoto)**

                    Progress Assessment

TRUE/FALSE Indicate whether each statement below is true or false by writing **T** or **F** in the space provided.

______ **1.** As in the United States, people in Japan speak many different languages.

______ **2.** Shintoists believe that everything in nature has a spirit, called a *kami.*

______ **3.** The chief obstacle to the reunification of Korea is the question of government.

______ **4.** One of North Korea's greatest problems is traffic congestion.

______ **5.** Korean is a difficult language to learn because it does not use an alphabet.

MATCHING In the space provided, write the letter of the term or person that matches each description. Some answers will not be used.

______ **1.** helped South Korea after World War II

______ **2.** ruler of North Korea after 1994

______ **3.** Japan's highest mountain

______ **4.** second largest city in Japan

______ **5.** ruler of North Korea until 1994

______ **6.** Japan's largest mountain range

______ **7.** the name of Japan's elected legislature

______ **8.** the capital of North Korea

______ **9.** helped North Korea after World War II

______ **10.** fee a country charges on exports and imports

a. Kim Il Sung

b. Fuji

c. Pyongyang

d. United States

e. tariff

f. Soviet Union

g. Kim Jong Il

h. Alps

i. Diet

j. Osaka

k. Honshu

l. Kyushu

Japan and the Koreas

Chapter Test

Form B

SHORT ANSWER Answer each of the following questions in complete sentences. Remember to use specific examples to support your answers.

1. How did Japan build a strong economy?

____

____

2. Why are some countries worried about North Korea?

____

3. Why is North Korea poorer than South Korea or Japan?

____

____

PRACTICING SOCIAL STUDIES SKILLS Study the map below and answer the question that follows.

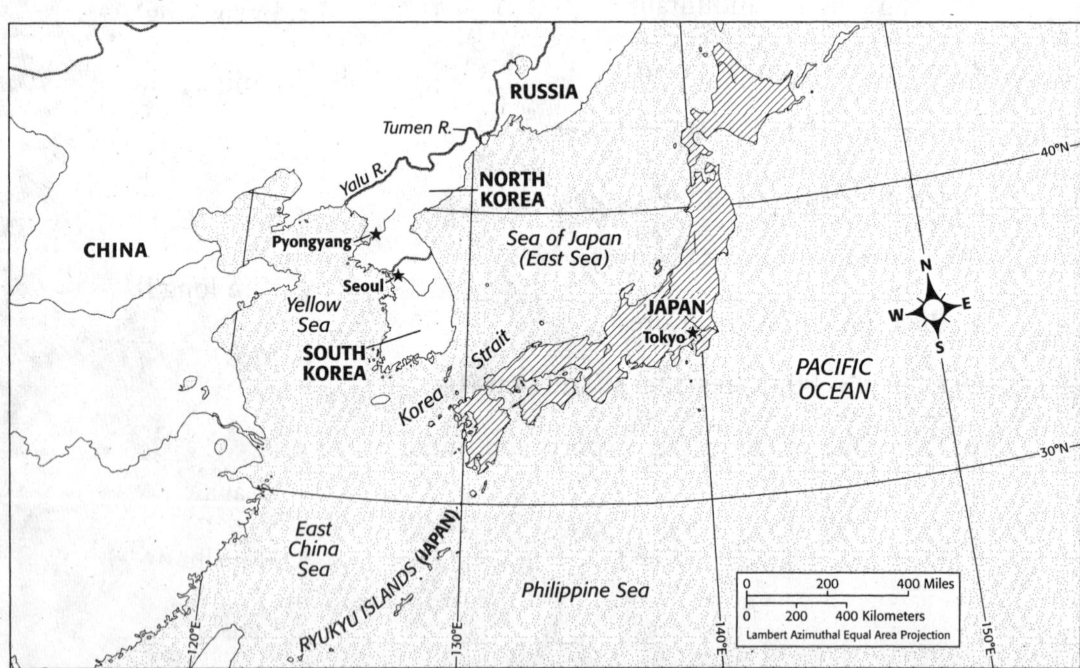

1. Japan was one of the most isolated countries in the world until the 1800s. Look at the map, and then explain why you think this might have happened.

____

____

  Progress Assessment

Southeast Asia

Section Quiz

Section 1

TRUE/FALSE Mark each statement **T** if it is true or **F** if it is false. If false explain why.

_______ **1.** Southeast Asia has two island groups, the Philippines and the Indochina Archipelago.

____

____

_______ **2.** The Mekong River is the region's most important river.

____

____

_______ **3.** A large group of islands is called a peninsula.

____

____

_______ **4.** The region's humid tropical climate helps support tropical rain forests.

____

____

_______ **5.** The region's climate and soil make farming very difficult.

____

____

Southeast Asia

Section Quiz

Section 2

MULTIPLE CHOICE For each of the following, write the letter of the best choice in the space provided.

______ **1.** The region's most advanced early civilization was the
  a. Thai.
  b. Malay.
  c. Khmer.
  d. Buddhist.

______ **2.** The Dutch once controlled the Dutch East Indies, which is now the country of
  a. India.
  b. Malaysia.
  c. Laos.
  d. Indonesia.

______ **3.** What were the three independent countries that Indochina split into after the French left?
  a. Timor, East Timor, Indonesia
  b. Malaysia, Brunei, Indonesia
  c. Borneo, New Guinea, Java
  d. Cambodia, Laos, Vietnam

______ **4.** Which European country spread Roman Catholicism to the Philippines?
  a. Spain
  b. Portugal
  c. France
  d. Great Britain

______ **5.** Buddhist temples that also serve as monasteries are called
  a. monks.
  b. sarongs.
  c. wats.
  d. sites.

Progress Assessment

Southeast Asia   Section Quiz

Section 3

FILL IN THE BLANK For each of the following statements, fill in the
blank with the appropriate word, phrase, or name.

1. Some countries around the world refuse to trade with Myanmar because its

   government abuses ___________________________.

2. The canals, or ___________________________, of Bangkok are used for travel, trade,
   and draining floodwater.

3. The countries of Mainland Southeast Asia include Myanmar, Thailand,

   Cambodia, Laos, and ___________________________.

4. The rapid growth of the cities of Island Southeast Asia has led to

   ___________________________ and pollution.

5. The capital and chief city of Cambodia is ___________________________.

Southeast Asia

Section Quiz

Section 4

MATCHING In the space provided, write the letter of the term or place that matches each description. Some answers will not be used.

_______ **1.** village or city district with traditional homes on stilts

_______ **2.** shipping center that places few or no taxes on goods

_______ **3.** the capital of the Philippines

_______ **4.** the main island of Indonesia

_______ **5.** country that includes the southern Malay Peninsula and northern Borneo

_______ **6.** supreme ruler of a Muslim country

_______ **7.** largest and most populated island of the Philippines

_______ **8.** largest island country of Southeast Asia

_______ **9.** capital of Malaysia

_______ **10.** tiny country on the island of Borneo

a. free port

b. Jakarta

c. Singapore

d. kampong

e. Malaysia

f. Java

g. Manila

h. Kuala Lumpur

i. sultan

j. Luzon

k. Brunei

l. Indonesia

Progress Assessment

Southeast Asia

Chapter Test

Form A

UNDERSTANDING MAIN IDEAS For each of the following, write the letter of the best choice in the space provided.

______ **1.** Mainland Southeast Asia has mainly a
   a. tropical savanna climate.
   b. highland climate.
   c. monsoon climate.
   d. steppe climate.

______ **2.** What is the capital of Myanmar?
   a. Bangkok
   b. Phnom Penh
   c. Hanoi
   d. Yangon

______ **3.** The supreme ruler of a Muslim country, such as Brunei, is a
   a. president.
   b. sultan.
   c. monk.
   d. monarch.

______ **4.** Which country did the United States grant independence to after World War II?
   a. Myanmar
   b. Thailand
   c. the Philippines
   d. Singapore

______ **5.** What are the two peninsulas of Southeast Asia?
   a. Philippine Peninsula and Thailand Peninsula
   b. Indochina Peninsula and Malay Peninsula
   c. Myanmar Peninsula and Malay Peninsula
   d. Indochina Peninsula and Vietnam Peninsula

______ **6.** The main religions of Southeast Asia are Buddhism, Christianity, Hinduism, and
   a. animism.
   b. Judaism.
   c. Islam.
   d. Jainism.

______ **7.** Although a few Filipinos are wealthy, most are poor
   a. fishers.
   b. factory workers.
   c. shop keepers.
   d. farmers.

______ **8.** Many of the plants and animals of Southeast Asia's tropical rain forests are endangered because of
   a. biodiversity.
   b. deforestation.
   c. flooding.
   d. tsunamis.

Progress Assessment

_____ **9.** Some countries will not trade with Myanmar because of its poor record on
  a. human rights.
  b. environmental issues.
  c. religious rights.
  d. economic issues.

_____ **10.** After Dutch traders ousted the Portuguese in the 1600s and 1700s, Portugal kept only the small island of
  a. Java.
  b. Borneo.
  c. Timor.
  d. Luzon.

_____ **11.** What is the main reason why Singapore is a rich country?
  a. It has lots of rich farmland.
  b. It is a major tourist center.
  c. It is located near the United States.
  d. It is located on a major shipping route.

_____ **12.** What causes tsunamis?
  a. monsoons and typhoons
  b. underwater earthquakes and volcanic eruptions
  c. melting glaciers
  d. deforestation and flooding

_____ **13.** What is the main religion in Indonesia?
  a. Islam
  b. Buddhism
  c. Hinduism
  d. Christianity

_____ **14.** Why is Laos considered undeveloped?
  a. It has few roads.
  b. It has no railroads.
  c. It has limited electricity.
  d. all of the above

_____ **15.** What is the main crop of Southeast Asia?
  a. corn
  b. wheat
  c. rice
  d. all of the above

_____ **16.** According to the domino theory,
  a. fighting for independence always results in a divided country.
  b. if one country fell to communism, its neighbors would also.
  c. the only way for a country to achieve peace is with the help of the United Nations.
  d. a monarch may serve as a head of state, but the legislature must hold the real power.

PRACTICING SOCIAL STUDIES SKILLS Study the map below and answer the question that follows.

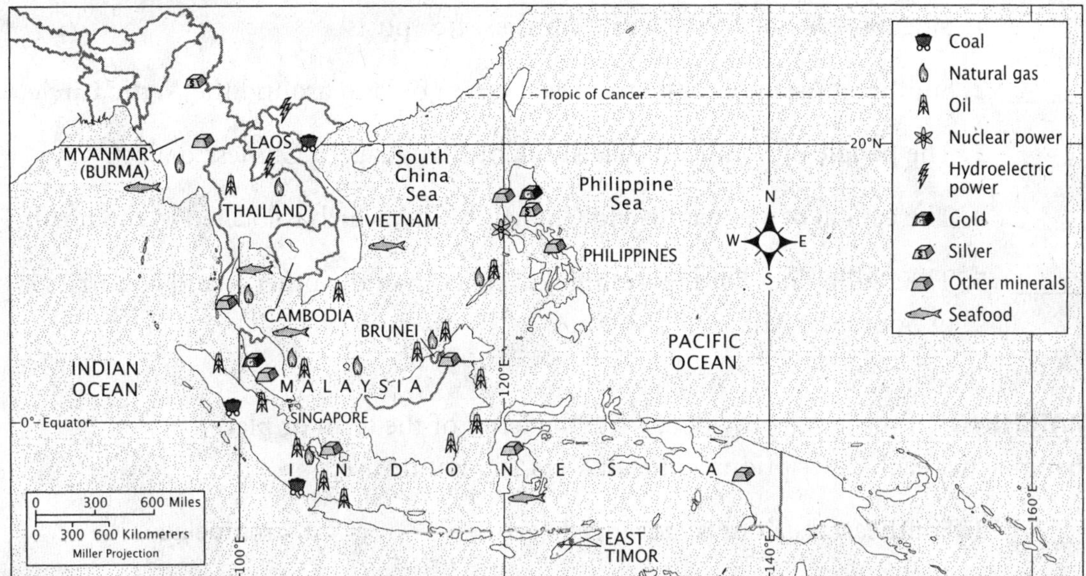

______ **1.** Which two countries have resources of gold?

    a. Indonesia and Malaysia

    b. Myanmar and Indonesia

    c. the Philippines and Malaysia

    d. the Philippines and Brunei

FILL IN THE BLANK Read each sentence and fill in the blank with the word in the word pair that best completes the sentence.

1. One part of Malaysia is on the Malay Peninsula and the other part is on

_______________________. **(Borneo/Java)**

2. Several major rivers drain Mainland Southeast Asia's peninsulas, including the

_______________________. **(Timor River/Mekong River)**

3. The United States sent troops to Vietnam in the 1960s to defend South Vietnam

from _______________________. **(Communist forces/European colonists)**

4. Most people of Southeast Asia live in _______________________.
(rural areas/cities)

5. The slums around Jakarta are called _______________________.
(kampongs/klongs)

TRUE/FALSE Indicate whether each statement below is true or false by writing **T** or **F** in the space provided.

_______ **1.** Singapore is one of the world's busiest free ports.

_______ **2.** Southeast Asia lies in the tropics, the area on and around the Arctic Circle.

_______ **3.** The people of Southeast Asia speak one language, a Chinese dialect.

_______ **4.** The capital of Vietnam is Ho Chi Minh City, which is located in the north.

_______ **5.** The Philippines have less ethnic diversity than the other island countries of Southeast Asia.

MATCHING In the space provided, write the letter of the term or place that matches each description. Some answers will not be used.

_______ **1.** Southeast Asia's most advanced early civilization

_______ **2.** a large group of islands

_______ **3.** island on which more than half of Indonesia's people live

_______ **4.** island that includes Malaysia, Brunei, and Indonesia

_______ **5.** Malaysia's capital, located on the Malay Peninsula

_______ **6.** huge storms that can bring heavy rains and strong winds

_______ **7.** giant series of waves

_______ **8.** villages or areas of cities with traditional homes built on stilts to protect against flooding

_______ **9.** the capital of Indonesia

_______ **10.** Bangkok's network of canals

a. archipelago

b. Borneo

c. typhoons

d. tsunamis

e. Jakarta

f. Khmer

g. wats

h. klongs

i. kampongs

j. Kuala Lumpur

k. Manila

l. Java

Southeast Asia

Chapter Test

Form B

SHORT ANSWER Answer each of the following questions in complete sentences. Remember to use specific examples to support your answers.

1. What is the climate of Southeast Asia like?

____

____

2. Why did European powers come to this region?

____

____

3. What problems have hurt Indonesia's economy?

____

____

PRACTICING SOCIAL STUDIES SKILLS Study the map below and answer the question that follows.

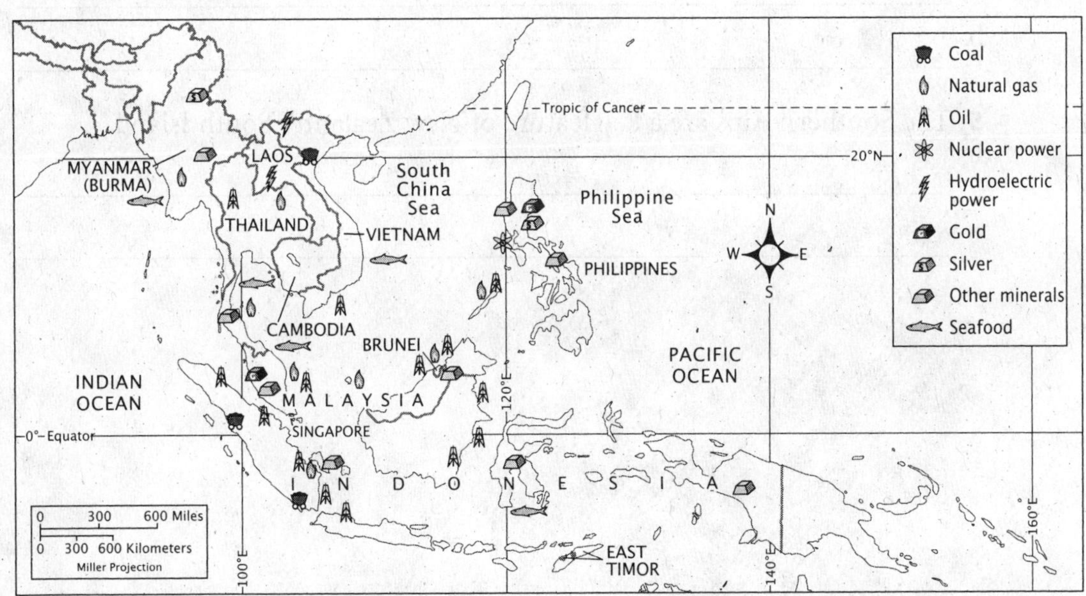

1. What are the differences and similarities between the resources of Vietnam and those of the Philippines?

____

____

                                   Progress Assessment

The Pacific World   Section Quiz

Section 1

TRUE/FALSE Mark each statement **T** if it is true or **F** if it is false. If false explain why.

______ **1.** Australia's Great Barrier Reef is the largest coral reef in the world.

____

____

______ **2.** The Maori came to Australia over 40,000 years ago.

____

____

______ **3.** Sydney is one of New Zealand's largest cities.

____

____

______ **4.** Mining is an important industry in the Outback, Australia's interior.

____

____

______ **5.** The Southern Alps are a key feature of New Zealand's South Island.

____

____

                    Progress Assessment

The Pacific World

Section Quiz

Section 2

FILL IN THE BLANK For each of the following statements, fill in the blank with the appropriate word, phrase, or name.

1. Of the three regions of the Pacific Islands, _______________________ is the largest.

2. Most _______________________ islands tend to be mountainous and rocky.

3. A _______________________ is an area that is under the authority of another government.

4. Fishing, tourism, and _______________________ are key industries in the Pacific Islands.

5. An _______________________ is a small, ring-shaped coral island that surrounds a lagoon.

Name _____________________________ Class _____________ Date _____________

The Pacific World

Section Quiz

Section 3

MULTIPLE CHOICE For each of the following, write the letter of the best choice in the space provided.

______ **1.** Antarctica is in almost total darkness in winter months because of
   a. its low latitude.
   b. its high latitude.
   c. the thinning ozone layer.
   d. its high elevation.

______ **2.** The Antarctic Peninsula was first sighted in 1775 by
   a. Sir Ernest Shackleton.
   b. James Cook.
   c. Roald Amundsen.
   d. Ferdinand Magellan.

______ **3.** Among others, which of the following countries currently hosts a research station in Antarctica?
   a. Fiji
   b. Afghanistan
   c. Russia
   d. Papua New Guinea

______ **4.** The ice sheet that covers Antarctica
   a. contains more than 90 percent of the world's ice.
   b. is about the size of France.
   c. is a few feet thick.
   d. has no vegetation at all.

______ **5.** An international agreement reached in 1991 protects Antarctica from
   a. mining.
   b. drilling.
   c. tourism.
   d. all of the above.

Progress Assessment

The Pacific World

Chapter Test

Form A

MULTIPLE CHOICE For each of the following, write the letter of the best choice in the space provided.

______ **1.** What protects Earth's living things from the harmful effects of the sun's ultraviolet rays?
 a. global warming
 b. the ozone layer
 c. Antarctic ice shelves
 d. the polar desert

______ **2.** What Pacific Island region includes New Guinea and Fiji?
 a. Cook Islands
 b. Melanesia
 c. Marshall Islands
 d. Polynesia

______ **3.** What type of climate does most of Antarctica have?
 a. highland
 b. marine
 c. steppe
 d. ice cap

______ **4.** The Southern Alps are a key feature of
 a. South Island.
 b. Tasmania.
 c. North Island.
 d. Kiribati.

______ **5.** An atoll is a
 a. volcanic island.
 b. coral island.
 c. sandstone island.
 d. continental island.

______ **6.** What did the Antarctic Treaty of 1959 ban?
 a. mining
 b. tourism
 c. scientific research
 d. military activity

______ **7.** Which of the following is a typical feature of low islands?
 a. coconut palms
 b. rugged mountains
 c. fertile soil
 d. dense forests

______ **8.** Who were the first humans to live in Australia?
 a. Maori
 b. European settlers
 c. Aborigines
 d. British prisoners

Progress Assessment

_______ 9. Ice shelves in Antarctica form
  a. on mountain tops.
  b. along the coast.
  c. in the interior valleys.
  d. on floating icebergs.

_______ 10. Australia and New Zealand
  are the top producers of
  a. rice.
  b. wool.
  c. oil.
  d. heavy machines.

_______ 11. Australia's two largest cities
  are
  a. Wellington and Auckland.
  b. Auckland and Sydney.
  c. Port Moresby and
     Melbourne.
  d. Sydney and Melbourne.

_______ 12. Papua New Guinea is the
  eastern part of a
  a. high island.
  b. coral island.
  c. low island.
  d. Polynesian island.

_______ 13. Which Pacific island region
  includes about 2,000 small
  islands?
  a. Polynesia
  b. Melanesia
  c. Micronesia
  d. Marshall Islands

_______ 14. Where do most of the people
  in New Zealand live?
  a. on South Island
  b. in Wellington
  c. in rural areas
  d. on North Island

_______ 15. Most Pacific islanders are
  a. European.
  b. Christian.
  c. Indian.
  d. Muslim.

_______ 16. The western half of Australia
  is covered by
  a. a huge plateau.
  b. low mountains.
  c. valleys.
  d. coral reefs.

_______ 17. Which countries controlled
  most of the Pacific Islands in
  the late 1800s?
  a. the United States, France,
     and New Zealand
  b. Germany, Japan, and the
     United States
  c. Spain, Great Britain, and
     France
  d. Great Britain, Japan, and
     Germany

PRACTICING SOCIAL STUDIES SKILLS Study the charts below and answer the question that follows.

Ethnic Groups in Australia and New Zealand

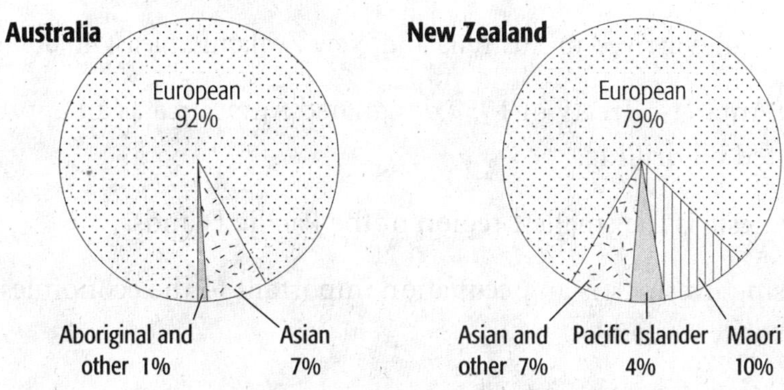

_______ **1.** According to the figures in the two pie charts,
   a. Aborigines are greater in number than ethnic Asians in Australia.
   b. Ethnic Europeans represent more of New Zealand's population than do Europeans in Australia.
   c. Maori are fewer in number than ethnic Asians in New Zealand.
   d. Aborigines represent less than one percent of Australia's population.

FILL IN THE BLANK Read each sentence and fill in the blank with the word in the word pair that best completes the sentence.

1. Masses of ice called _______________________ float in the waters surrounding Antarctica. (**icebergs/glaciers**)

2. Mining is an important industry in the _______________________, Australia's interior. (**Great Barrier Reef/Outback**)

3. In _______________________, Indians make up nearly half of the population. (**Fiji/North Island**)

4. A _______________________ is a high-latitude region that receives little precipitation. (**polar desert/marine**)

5. The _______________________ came from other Pacific islands to settle New Zealand. (**Aborigines/Maori**)

TRUE/FALSE Indicate whether each statement below is true or false by writing **T** or **F** in the space provided.

_______ **1.** The islands in Micronesia are east of the Philippines.

_______ **2.** Aborigines arrived in Australia and New Zealand at least 40,000 years ago.

_______ **3.** The Antarctic Treaty of 1959 designated Antarctica as an approved site for limited oil drilling.

_______ **4.** Polynesia is the smallest region of the Pacific Islands.

_______ **5.** Raising livestock is an occupation important to the economies of Australia and New Zealand.

MATCHING In the space provided, write the letter of the term or place that matches each description. Some answers will not be used.

_______ **1.** mountain range in New Zealand   **a.** Uluru

_______ **2.** native Australian animal   **b.** coral reef

_______ **3.** most heavily populated Pacific Islands region   **c.** Melbourne

_______ **4.** an area that is under the authority of another government   **d.** koala

_______ **5.** chain of rocky materials found in shallow tropical waters   **e.** Melanesia

     f. kiwi

_______ **6.** Pacific Islands region that includes Tahiti and Hawaii   **g.** territory

_______ **7.** native New Zealand bird   **h.** air pollution

_______ **8.** a rock formation in western Australia   **i.** Auckland

_______ **9.** large city in New Zealand   **j.** Polynesia

_______ **10.** the likely cause of the thinning of Earth's ozone layer   **k.** Southern Alps

     l. Tonga

                    Progress Assessment

The Pacific World

Chapter Test
Form B

SHORT ANSWER Answer each of the following questions in complete
sentences. Remember to use specific examples to support your answer.

1. In what ways are Australia and New Zealand different from one another?

____

____

2. Describe three activities that have threatened Antarctica's environment.

____

____

3. Describe the three Pacific Island regions.

____

____

PRACTICING SOCIAL STUDIES SKILLS Study the charts below and
answer the question that follows.

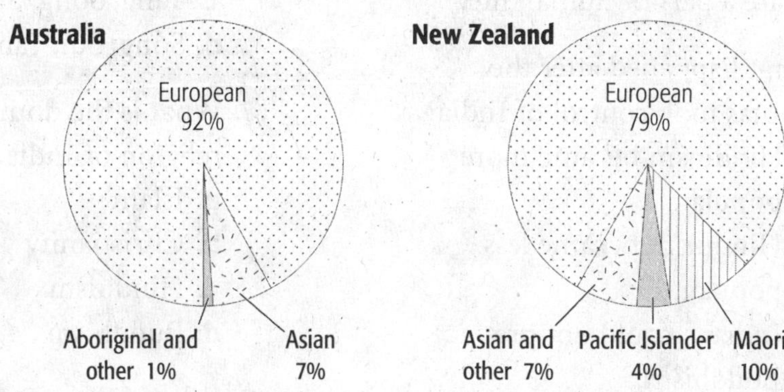

1. Compare the Aboriginal population of Australia to the Maori population of
New Zealand. Based on these figures and your knowledge about Australia and
New Zealand, what conclusions can you draw about the difference in population?

____

____

South and East Asia and the Pacific

Benchmark Test

Form A

MULTIPLE CHOICE For each of the following, write the letter of the best choice in the space provided.

______ **1.** Both Harappa and Mohenjo Daro were located near the
- **a.** Arabian Sea.
- **b.** Thar Desert.
- **c.** Chang Jiang River.
- **d.** Indus River.

______ **2.** With which of the following statements would the Buddha likely disagree?
- **a.** Suffering comes from not having what one wants.
- **b.** Nirvana is achieved by overcoming ignorance and desire.
- **c.** Contentment springs from gaining what one wants.
- **d.** Suffering and unhappiness are a part of human life.

______ **3.** What happened after the Guptas took control of India?
- **a.** Hinduism became more popular.
- **b.** Hinduism became less popular.
- **c.** Indians could not practice Buddhism.
- **d.** Indians could not practice Jainism.

______ **4.** The dynasty during which the Mongols ruled China was the
- **a.** Khan.
- **b.** Yuan.
- **c.** Song.
- **d.** Han.

______ **5.** Which of the following best describes the Qin dynasty under Shi Huangdi?
- **a.** His policies led to rebellion and civil war.
- **b.** His weak leadership resulted in the Warring States period.
- **c.** China lost territory under his rule.
- **d.** His strict policies kept China unified.

______ **6.** Which of the following correctly shows the order of dynasties in China?
- **a.** Sui, Song, Tang
- **b.** Sui, Tang, Song
- **c.** Tang, Song, Sui
- **d.** Song, Sui, Tang

______ **7.** What is the dominant religion of India today?
- **a.** Islam
- **b.** Christianity
- **c.** Hinduism
- **d.** Buddhism

______ **8.** What did the British agree to in 1947?
- **a.** independence for Bangladesh
- **b.** freeing Ghandi from prison
- **c.** ending the caste system
- **d.** the partition of India

______ **9.** For much of the 20th century Mongolia was under the influence of
   a. Japan.
   b. Taiwan.
   c. the Soviet Union.
   d. China.

______ **10.** Most Chinese live in the
   a. Taklimakan Desert.
   b. North China Plain.
   c. Plateau of Tibet.
   d. Sichuan Basin.

______ **11.** What religion or belief system was brought to Korea and later carried to Japan by Chinese missionaries?
   a. Shinto
   b. Buddhism
   c. Confucianism
   d. Christianity

______ **12.** Compared to California, Japan is
   a. half as large, with twice the population.
   b. twice as large, with half the population.
   c. about the same size, with four times the population.
   d. four times as large, with about the same population.

______ **13.** What causes tsunamis?
   a. monsoons and typhoons
   b. underwater earthquakes and volcanic eruptions
   c. melting glaciers
   d. deforestation and flooding

______ **14.** According to the domino theory
   a. if one country fell to communism, its neighbors would also.
   b. fighting for independence always results in a divided country.
   c. the only way for a country to achieve peace is with the help of the United Nations.
   d. a monarch may serve as head of state, but the legislature must hold the real power.

______ **15.** What type of climate does most of Antarctica have?
   a. ice cap
   b. marine
   c. steppe
   d. highland

______ **16.** Who were the first humans to live in Australia?
   a. Aborigines
   b. Maori
   c. Dutch settlers
   d. British prisoners

PRACTICING SOCIAL STUDIES SKILLS Study the map below and
answer the question that follows.

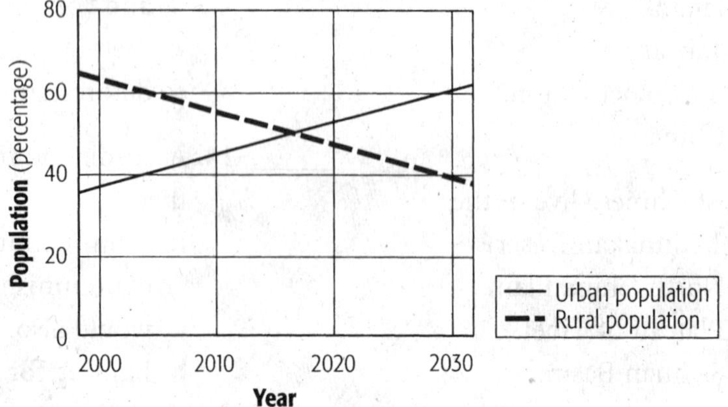

______ **1.** According to the graph, in about what year will the percentage of China's
        urban and rural population be the same?
        a. 2009
        b. 2017
        c. 2023
        d. 2030

FILL IN THE BLANK Read each sentence and fill in the blank with the
word in the word pair that best completes the sentence.

1. The ______________________ arrived in the Indus River Valley from the
    northwest. (**Harrapans/Aryans**)

2. Gupta society ______________________ after the rule of Candra Gupta II.
    (**began to decline/adopted Hinduism**)

3. ______________________ taught proper behavior, and also emphasized spiritual
    matters. (**Confucianism/Neo-Confucianism**)

4. Leaders of the Song dynasty used the ______________________ examination to
    make sure qualified people became government workers.
    (**scholar officials/civil service**)

5. ______________________ are seasonal winds that bring either moist or dry air
    to an area. (**Savannas/Monsoons**)

6. The _____________________ was a program in India to encourage farmers to adopt modern farming methods. (**green revolution/partition**)

7. _____________________ is green and tropical, with beautiful mountains and crowded cities. (**Mongolia/Taiwan**)

8. The _____________________ was built near China's northern border to stop invaders from the north. (**Three Gorges Dam/Great Wall**)

9. During World War II, the United States dropped atomic bombs on Hiroshima and _____________________. (**Kyoto/Nagasaki**)

10. Most people of Southeast Asia live in _____________________. (**cities/rural areas**)

11. Mining is an important industry in the _____________________, Australia's interior. (**Great Barrier Reef/Outback**)

12. A _____________________ is a high latitude region that receives little precipitation. (**polar desert/marine**)

TRUE/FALSE Indicate whether each statement below is true or false by writing **T** or **F** in the space provided.

_______ **1.** The caste system divided Indian society into groups based on a person's birth, wealth, or occupation.

_______ **2.** During the Zhou dynasty, people believed their rulers had a mandate of heaven.

_______ **3.** Mohandas Ghandi was a main opponent of the Indian independence movement.

_______ **4.** Millions of people in India have moved out of cities in search of jobs.

_______ **5.** Taiwan is an island nation, while Mongolia is landlocked.

_______ **6.** China welcomed contact with the outside world throughout much of its history.

_______ **7.** Korea's main obstacle to reunification is the question of a common form of government.

_______ **8.** Japan has many diverse ethnic groups that all speak different languages.

______ **9.** Southeast Asia lies in the tropics, the area around the Arctic Circle.

______ **10.** Singapore is one of the world's busiest free ports.

______ **11.** The Antarctic Treaty of 1959 designated the continent as an approved site for oil drilling and military activity.

______ **12.** Raising livestock is important to the economies of Australia and New Zealand.

MATCHING In the space provided, write the letter of the term or place that matches each description. Some answers will not be used.

______ **1.** people who want to spread their religious beliefs

______ **2.** most heavily populated Pacific Island region

______ **3.** a landform at the mouth of a river created by sediment deposits

______ **4.** the most important language of ancient India

______ **5.** a group of nomadic peoples who invaded China

______ **6.** a large group of islands

______ **7.** the world's coldest desert

______ **8.** fees a country charges on exports and imports

______ **9.** territory claimed by both India and Pakistan

______ **10.** Japan's highest mountain

a. Gobi

b. Melanesia

c. Mongols

d. wats

e. delta

f. Kashmir

g. tariffs

h. Polynesia

i. missionaries

j. archipelago

k. Sanskrit

l. Fuji

South and East Asia and the Pacific

Benchmark Test

Form B

SHORT ANSWER Answer each of the following questions in complete sentences. Remember to use specific examples to support your answers.

1. What were some of the major accomplishments of the Gupta period?

____

____

2. Why was the Tang dynasty called the golden age of Chinese civilization?

____

____

3. Describe the climate of the Indian Subcontinent.

____

____

4. Why is Mongolia sparsely populated?

____

____

5. How did Japan build a strong economy?

____

____

6. Why did European powers come to Southeast Asia?

____

____

7. What activities have threatened Antarctica's environment?

____

____

8. Describe the three Pacific Island regions.

____

____

PRACTICING SOCIAL STUDIES SKILLS Study the map below and
answer the questions that follow.

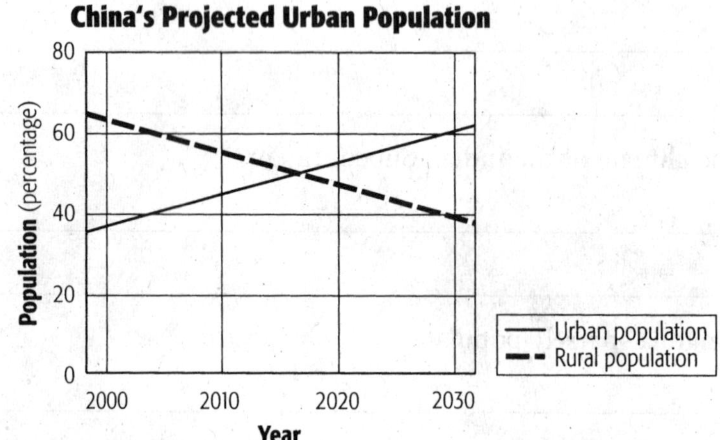

1. According to the graph, by about how much will the percentage of China's rural
population decrease between 2000 and 2030?

____

____

2. How might you explain the changes in China's population as shown on the graph?

____

____

Holt Social Studies

South and East Asia and the Pacific

South and East Asia and the Pacific

DIAGNOSTIC TEST

Multiple Choice

1. d	**9.** c
2. c	**10.** b
3. a	**11.** b
4. b	**12.** c
5. d	**13.** b
6. b	**14.** a
7. c	**15.** a
8. d	**16.** a

Practicing Social Studies Skills

1. b

Fill in the Blank

1. Aryans	**7.** Taiwan
2. began to decline	**8.** Great Wall
3. Neo-Confucianism	**9.** Nagasaki
4. civil service	**10.** rural areas
5. Monsoons	**11.** Outback
6. green revolution	**12.** polar desert

True/False

1. T	**7.** T
2. T	**8.** F
3. F	**9.** F
4. F	**10.** T
5. T	**11.** F
6. F	**12.** T

Matching

1. i	**6.** j
2. b	**7.** a
3. e	**8.** g
4. k	**9.** f
5. c	**10.** l

History of Ancient India

SECTION QUIZ

Section 1

1. c	**4.** a
2. a	**5.** d
3. b	

Section 2

1. Aryans	**6.** Reincarnation
2. Brahmanism	**7.** Brahman
3. *Rigveda*	**8.** Karma
4. caste system	**9.** Jainism
5. Hinduism	**10.** nonviolence

Section 3

1. l	**6.** j
2. d	**7.** c
3. h	**8.** i
4. m	**9.** n
5. e	**10.** g

Section 4

1. F; In the 320s BC a military leader named Candragupta Maurya seized control of the entire northern part of India.

2. T

3. T

4. F; Rulers of the Gupta dynasty believed the strict social order of the Hindu caste system would strengthen their rule.

5. F; India was firmly under Gupta control until the late 400s, when the Huns invaded India from the northwest.

Section 5

1. religious	**6.** *Bhagavad Gita*
2. Sanskrit	**7.** temples
3. metallurgy	**8.** alloy
4. inoculation	**9.** Indian
5. Hindu	**10.** astronomy

CHAPTER TEST FORM A

Multiple Choice

1. d	**7.** b
2. d	**8.** a
3. c	**9.** a
4. a	**10.** d
5. c	**11.** a
6. b	

Practicing Social Studies Skills

1. c

Fill in the Blank

1. Aryans

2. Jainism

3. caste

4. Candra Gupta II

5. alloys

True/False

1. F	**4.** F
2. T	**5.** F
3. T	

Matching

1. i	**6.** g
2. a	**7.** h
3. c	**8.** f
4. d	**9.** e
5. k	**10.** b

CHAPTER TEST FORM B
Short Answer

1. People used irrigation canals to bring water from the Indus river to their fields. Cities were located near the river.
2. The Buddha believed that suffering came from people desiring material goods that they did not have, and that they should learn to overcome this desire.
3. During the Gupta period, architecture became more complex, great paintings and sculptures were created, and new ideas in math and medicine were developed.

Practicing Social Studies Skills

1. There are probably more Buddhists in India, because it was closer to the area where Buddhism started.

History of Ancient China
SECTION QUIZ
Section 1

1. F; Land along the Huang He and Chang Jiang was good for farming.
2. T
3. F; The Zhou dynasty lasted longer than any other dynasty in Chinese history.
4. T
5. T

Section 2

1. Wudi	**6.** father
2. peasants	**7.** seismograph
3. paper	**8.** earthquakes
4. Confucius	**9.** sundial
5. Liu Bang	**10.** Acupuncture

Section 3

1. i	**6.** j
2. b	**7.** l
3. g	**8.** n
4. m	**9.** h
5. a	**10.** c

Section 4

1. Confucianism
2. ethics
3. order
4. Han
5. Buddhism
6. bureaucracy
7. Civil service
8. scholar-official
9. Neo-Confucianism
10. Song

Section 5

1. k	**6.** n
2. d	**7.** c
3. h	**8.** m
4. j	**9.** f
5. i	**10.** e

CHAPTER TEST FORM A
Multiple Choice

1. b	**6.** b
2. d	**7.** c
3. d	**8.** a
4. d	**9.** b
5. c	**10.** d

Practicing Social Studies Skills

1. c

Fill in the Blank

1. writing
2. Neo-Confucianism
3. Great Wall
4. daily activities
5. civil service

True/False

1. T	**4.** F
2. T	**5.** F
3. F	

Matching

1. d	**6.** a
2. i	**7.** e
3. l	**8.** m
4. c	**9.** f
5. b	**10.** h

CHAPTER TEST FORM B
Short Answer

1. The Tang dynasty produced great rulers who expanded Chinese territory and encouraged art and culture to grow.
2. All foreign influences were eliminated from Chinese society. People who challenged government authority were punished.
3. Wudi wanted a strong central government with a policy of Confucianism, where people got jobs based on ability.

Practicing Social Studies Skills

1. Major cities were located near the Huang He river and could have been destroyed by major floods.

The Indian Subcontinent

SECTION QUIZ
Section 1

1. F; The Ganges River creates a huge delta in Bangladesh.
2. F; The Himalayas in the north have a cool highland climate.
3. T
4. T
5. F; Monsoons are seasonal winds that bring either moist or dry air to a region.

Section 2

1. j	**6.** g
2. d	**7.** a
3. i	**8.** c
4. l	**9.** b
5. f	**10.** k

Section 3

1. c	**4.** b
2. d	**5.** d
3. a	

Section 4

1. Nepal	**4.** Kashmir
2. Bangladesh	**5.** Sherpas
3. tsunami	

CHAPTER TEST FORM A
Multiple Choice

1. b	**11.** a
2. d	**12.** d
3. b	**13.** d
4. c	**14.** b
5. d	**15.** a
6. c	**16.** b
7. c	**17.** b
8. d	**18.** d
9. b	**19.** c
10. a	**20.** d

Practicing Social Studies Skills

1. c

Fill in the Blank

1. Indus	**4.** Monsoons
2. Kathmandu	**5.** green revolution
3. Aryans	

True/False

1. F	**4.** T
2. F	**5.** F
3. T	

Matching

1. e	**6.** b
2. i	**7.** l
3. h	**8.** d
4. g	**9.** f
5. j	**10.** a

CHAPTER TEST FORM B
Short Answer

Answers will vary. Possible responses are provided.

1. Britain agreed to partition India into Hindu- and Muslim-dominated nations. As a result, many people moved to live in the country where their religion was in the majority.
2. Tropical climates—tropical savanna and humid tropical—dominate much of the subcontinent. Generally, it is warm with wet and dry seasons.

Progress Assessment

3. India faces several challenges. Rapid population growth strains its environment and resources. Reducing poverty and resolving its conflict with Pakistan over Kashmir are also challenges.

Practicing Social Studies Skills

1. possible answer—The northern part of the region has two large river valleys that provided fertile land for agriculture and could support a constant population.

China, Mongolia, and Taiwan

SECTION QUIZ
Section 1

1. b	**6.** l
2. h	**7.** f
3. a	**8.** i
4. c	**9.** g
5. e	**10.** d

Section 2

1. Han	**6.** dialect
2. Taiwan	**7.** calligraphy
3. dynasty	**8.** Daoism
4. Mao Zedong	**9.** North China
5. Confucianism	**10.** Pagodas

Section 3

1. b	**4.** d
2. a	**5.** b
3. c	

Section 4

1. F; The Nationalists established military rule; China had established a Communist government.
2. T
3. T
4. F; Mongolia was a Communist country until after the fall of the Soviet Union.
5. F; Ulaanbaatar is Mongolia's only large city.

CHAPTER TEST A
Multiple Choice

1. a	**9.** c
2. b	**10.** c
3. b	**11.** a
4. a	**12.** b
5. c	**13.** d
6. a	**14.** d
7. b	**15.** b
8. a	**16.** d

Practice Social Studies Skills

1. c

Fill in the Blank

1. Three Gorges Dam
2. Great Wall
3. Ulaanbaatar
4. Taiwan
5. Daoism

True/False

1. T	**4.** F
2. F	**5.** T
3. T	

Matching

1. e	**6.** j
2. f	**7.** i
3. b	**8.** c
4. g	**9.** h
5. a	**10.** d

CHAPTER TEST B
Short Answer

1. The government has responded harshly. It stopped the pro-democracy protestors in Beijing's Tiananmen Square in 1989, and it crushed a rebellion in Tibet in 1959.
2. Much of Mongolia is desert or grassland; there are shortages of both food and water.
3. In the southeast, the climate is generally warm and hot, with heavy rains. In the northwest, it is dry and temperatures vary.

Practicing Social Studies Skills

1. The percentage of China's urban population will rise by more than 20 percent between 2000 and 2030.

Progress Assessment

Japan and the Koreas

SECTION QUIZ

Section 1
1. T
2. F; Underwater earthquakes sometimes cause destructive waves called tsunamis.
3. F; Mountains cover about 75 percent of Japan.
4. T
5. T

Section 2
1. i
2. h
3. c
4. d
5. a
6. k
7. l
8. b
9. j
10. g

Section 3
1. b
2. c
3. d
4. b
5. a

Section 4
1. North
2. demilitarized zone
3. Republic
4. Communist
5. government
6. Kim Jong Il
7. strongest
8. corruption
9. nuclear
10. Seoul

CHAPTER TEST FORM A
Multiple Choice
1. b
2. a
3. c
4. d
5. b
6. b
7. a
8. c
9. d
10. a
11. c
12. c
13. b
14. d
15. d
16. a
17. d
18. b

Practicing Social Studies Skills
1. b

Fill in the Blank
1. China
2. Nagasaki
3. samurai
4. North Korea
5. Kyoto

True/False
1. F
2. T
3. T
4. F
5. F

Matching
1. d
2. g
3. b
4. j
5. a
6. h
7. i
8. c
9. f
10. e

CHAPTER TEST FORM B
Short Answer
1. Japan used a well-trained, loyal work force and efficient production techniques to produce high-quality manufactured goods for export.
2. North Korea is an isolated country that has announced that is has recently built nuclear weapons.
3. Although rich in mineral resources, North Korea uses many resources to make machinery or military supplies. Most factories are not efficient and are using out of date technology. It has also closed its markets to foreign goods.

Practicing Social Studies Skills
1. Japan is the most easterly Asian nation, far from the developing countries of the 1800s. Also, Japan is an island nation, which increases its ability to keep foreigners away.

Southeast Asia

SECTION QUIZ

Section 1
1. F; The two island groups are the Philippines and the Malay Archipelago.
2. T
3. F; A large group of islands is called an archipelago.
4. T
5. F; The region's climate and soil make farming very productive.

Section 2

1. c
2. d
3. d
4. a
5. c

Section 3

1. human rights
2. klongs
3. Vietnam
4. overcrowding
5. Phnom Penh

Section 4

1. d
2. a
3. g
4. f
5. e
6. i
7. j
8. l
9. h
10. k

CHAPTER TEST FORM A
Multiple Choice

1. a
2. d
3. b
4. c
5. b
6. c
7. d
8. b
9. a
10. c
11. d
12. b
13. a
14. d
15. c
16. b

Practicing Social Studies Skills

1. c

Fill in the Blank

1. Borneo
2. Mekong River
3. Communist forces
4. rural areas
5. kampongs

True/False

1. T
2. F
3. F
4. F
5. T

Matching

1. f
2. a
3. l
4. b
5. j
6. c
7. d
8. i
9. e
10. h

CHAPTER TEST FORM B
Short Answer

1. Temperatures are generally warm to hot all year. Much of the mainland has a tropical savanna climate with rainy summers and dry winters. The islands and Malay Peninsula have mainly a humid tropical climate that is hot, muggy, and rainy all year.
2. European powers came in search of spices and other trade goods, and to colonize and spread Christianity.
3. Indonesia's economy has been hurt by poverty, high unemployment, and religious and ethnic conflicts.

Practicing Social Studies Skills

1. The Philippines has gold, silver, other mineral resources, natural gas, and nuclear power. Vietnam has coal, hydroelectric power, and fishing. Both countries have oil resources.

The Pacific World

SECTION QUIZ
Section 1

1. T
2. F; The Aborigines came to Australia about 40,000 years ago.
3. F; Sydney is one of Australia's largest cities.
4. T
5. T

Section 2

1. Polynesia
2. high
3. territory
4. agriculture
5. atoll

Section 3

1. b
2. b
3. c
4. a
5. d

Progress Assessment

CHAPTER TEST FORM A
Multiple Choice

1. b	**10.** b
2. b	**11.** d
3. d	**12.** a
4. a	**13.** c
5. b	**14.** d
6. d	**15.** b
7. a	**16.** a
8. c	**17.** c
9. b	

Practicing Social Studies Skills
1. d

Fill in the Blank

1. icebergs	**4.** polar desert
2. Outback	**5.** Maori
3. Fiji	

True/False

1. T	**4.** F
2. F	**5.** T
3. F	

Matching

1. k	**6.** j
2. d	**7.** f
3. e	**8.** a
4. g	**9.** i
5. b	**10.** h

CHAPTER TEST FORM B
Short Answer

Answers will vary. Possible responses
are provided.

1. Australia is dry, hot, and flat; it has poor
soil and little fresh water. New Zealand is
mild, has mountains, dense forests, deep
lakes, and fertile soil. The first people in
Australia were Aborigines. The Maori were
the first people in New Zealand. Mining,
steel, heavy machines, and computers
are important industries in Australia.
Manufacturing, banking, insurance, and
tourism are important in New Zealand.

2. Tourism, oil spills, and mining have
threatened Antarctica's environment.

3. Micronesia is a group of about 2,000 small
islands located east of the Philippines.
Melanesia stretches from New Guinea to
Fiji and is the most populated. Polynesia is
the largest region and includes the islands
of Tonga, Samoa, and Hawaii.

Practicing Social Studies Skills
1. The Maori in New Zealand have
survived in far greater numbers than the
Aborigines in Australia. The nomadic
ways of the Aborigines may have hindered
their ability to survive after the British
settlers took over most of their land. On
the other hand, farming may have allowed
Maori groups to adapt to new places when
the British displaced them from their
traditional homes.

South and East Asia and the Pacific

BENCHMARK TEST FORM A
Multiple Choice

1. d	**9.** c
2. c	**10.** b
3. a	**11.** b
4. b	**12.** c
5. d	**13.** b
6. b	**14.** a
7. c	**15.** a
8. d	**16.** a

Practicing Social Studies Skills
1. b

Fill in the Blank

1. Aryans	**7.** Taiwan
2. began to decline	**8.** Great Wall
3. Neo-Confucianism	**9.** Nagasaki
4. civil service	**10.** rural areas
5. Monsoons	**11.** Outback
6. green revolution	**12.** polar desert

True/False

1. T	**7.** T
2. T	**8.** F
3. F	**9.** F
4. F	**10.** T
5. T	**11.** F
6. F	**12.** T

Progress Assessment

Answer Key

Matching

1. i		**6.** j	
2. b		**7.** a	
3. e		**8.** g	
4. k		**9.** f	
5. c		**10.** l	

BENCHMARK TEST FORM B

Short Answer

Answers will vary. Possible responses are provided.

1. Architecture became more complex, great paintings and sculptures were created, and new ideas in math and medicine developed.
2. The Tang dynasty produced great rulers who expanded Chinese territory and encouraged art and culture.
3. Tropical climates—tropical savanna and humid tropical—dominate much of the subcontinent. It is generally warm with wet and dry seasons.
4. Much of Mongolia is desert or grassland; there are shortages of both food and water.
5. Japan used a well-trained, loyal work force and efficient production techniques to produce high-quality manufactured goods for export.
6. They came in search of spices and other trade goods, to establish colonies, and to spread Christianity.
7. Tourism, mining, and oil spills have threatened Antarctica's environment.
8. Micronesia is a group of about 2,000 small islands located east of the Philippines. Melanesia stretches from New Guinea to Fiji and is the most populated region. Polynesia is the largest region and includes Tonga, Samoa, and Hawaii.

Practicing Social Studies Skills

1. China's rural population will decrease by about 20 percent between 2000 and 2030.
2. Many people may be moving from China's rural areas to its cities in search of jobs.

Progress Assessment